Guidelines for Preparation of State Water-Use Estimates for 2000

Edited by Joan F. Kenny

U.S. Department of the Interior
U.S. Geological Survey

U.S. Department of the Interior
Gale A. Norton, Secretary

U.S. Geological Survey
Charles G. Groat, Director

U.S. Geological Survey, Reston, Virginia: 2004

For sale by U.S. Geological Survey, Information Services
Box 25286, Denver Federal Center
Denver, CO 80225

For more information about the USGS and its products:
Telephone: 1-888-ASK-USGS
World Wide Web: http://www.usgs.gov/

Suggested citation:
Kenny, J.F., ed., 2004, Guidelines for preparation of State water-use estimates for 2000: U.S. Geological Survey Techniques and Methods 4–A4, 49 p., available on the World Wide Web at URL http://water.usgs.gov/watuse/

Prepared by the U.S. Geological Survey in Lawrence, Kansas (http://ks.water.usgs.gov)

Contents

Tables

Conversion Factors and Abbreviations

Multiply	By	To obtain
acre	43,560	square feet (ft^2)
acre	4,047	square meter (m^2)
acre-foot (acre-ft)	1,233	cubic meter (m^3)
acre-foot per day (acre-ft/d)	0.3259	million gallons per day (Mgal/d)
acare-foot per acre (acre-ft/acre)	3,047	cubic meter per hectare (m^3/ha)
foot (ft)	0.3048	meter (m)
gallon (gal)	3.785	liter (L)
gallon per capita per day (gpcd)	3.785	liter per capita per day (Lpcd)
gigawatt-hour (GWh)	1,000	kilowatt-hour (kWh)
kilowatt-hour (kWh)	3,412	British thermal unit (Btu)
megawatt	56,920	British thermal unit per minute (Btu/min)
megawatt	1,000,000	watts
mile (mi)	1.609	kilometer (km)
milligram per liter (mg/L)	1.0	parts per million (ppm)
million gallons (Mgal)	3.0689	acre-foot (acre-ft)
million gallons (Mgal)	3,785	cubic meter (m^3)
million gallons per day (Mgal/d)	1.3815	million cubic meter per year (Mm^3/yr)
million gallons per day (Mgal/d)	1.121	thousand acre-feet per year (kacre-ft/yr)
million gallons per ton (Mgal/ton)	4,172	cubic meter per megagram (m^3/Mg)
pound (lb)	453.6	gram (g)

Glossary

A

animal specialties water use Category in previous water-use compilations that included water use associated with fish farming and the raising of horses and fur-bearing animals such as rabbits and pets. For 2000, fish-farming water use is included with the aquaculture category, and water used for horses and other animals is included in the livestock category. *See also* aquaculture water use and livestock water use.

aquaculture water use Water used in the production of organisms that live in water within a confined space and under controlled feeding, sanitation, and harvesting procedures, and establishments primarily engaged in hatching fish and in operating fishing preserves. *See also* animal specialties water use, commercial water use, fish farms, and fish hatcheries.

aquifer A geologic formation, group of formations, or part of a formation that contains sufficient saturated permeable material to yield significant quantities of water to wells and springs.

B

blowdown The continuous or intermittent discharge, or purging, of a small amount of circulating water, such as in a boiler. Blowdown normally is expressed as a percentage of the water being circulated. Its purpose is to prevent an increase in the concentration of solids in the water due to evaporation.

C

capacity The average amount of water circulating in the cooling system of a thermoelectric powerplant, usually expressed in gallons per minute.

closed-loop cooling system A cooling system in which water is withdrawn, circulated through heat exchangers, then cooled and recycled. Subsequent withdrawals are used to replace water lost to evaporation, blowdown, drift, and leakage.

commercial water use Water used for motels, hotels, restaurants, office buildings, other commercial facilities, and institutions. The water may be obtained from a public supply or may be self supplied. In previous compilations, commercial water use included water use by fish hatcheries. *See also* aquaculture water use, public-supply water use, and self-supplied water.

consumptive use The part of water withdrawn that is evaporated, transpired, incorporated into products or crops, consumed by humans or livestock, or otherwise removed from the immediate water environment. Also referred to as water consumed.

conveyance loss Water that is lost due to leakage or evaporation while in transit through a pipe, canal, conduit, or ditch. Leakage from an irrigation ditch may percolate to a groundwater source and be available for further use.

cooling system Equipment that is used for cooling purposes, such as condensers at powerplants or factories. Includes water intakes and outlets, cooling towers, and cooling ponds.

cooling-system type *See* closed-loop cooling system and once-through cooling system.

D

deliveries Water distributed by public water suppliers for domestic, commercial, industrial, or thermoelectric uses.

dewatering The removal of water through draining or pumping to lower the water table for mining or agriculture.

dissolved solids A measure of the dissolved minerals and organic matter in water, usually expressed in milligrams per liter (mg/L). Water containing 1,000 mg/L or more dissolved solids is considered saline in this report.

domestic water use Water used for household purposes, such as drinking, food preparation, bathing, washing clothes and dishes, flushing toilets, and watering lawns and gardens. Also called residential water use. The water may be obtained from a public supply or may be self supplied. *See also* public-supply water use and self-supplied water.

drift Fine water droplets blown out of a cooling tower along with exhaust air, usually expressed as a percentage of water circulated.

E

evaporation Process by which water is changed from a liquid into a vapor. *See also* evapotranspiration and transpiration.

evapotranspiration Water that is vaporized as a result of evaporation from the soil or plant transpiration. *See also* evaporation and transpiration.

F

fish farms Facilities that produce finfish or shellfish under controlled feeding, sanitation, and harvesting procedures for commercial purposes. Water use by fish farms is included in the aquaculture category. *See also* animal specialties water use, aquaculture water use, and fish hatcheries.

fish hatcheries Facilities that raise fish for later release. Water use by fish hatcheries is included in the aquaculture category. *See also* aquaculture water use, commercial water use, and fish farms.

freshwater Water that contains less than 1,000 mg/L dissolved solids. Generally, water with more than 500 mg/L dissolved solids is undesirable for drinking and many industrial uses. *See also* saline water.

G

gigawatt-hour (GWh) A unit of energy equivalent to 1 billion watt-hours.

ground water All subsurface water, distinct from surface water. Specifically, that part of the subsurface water in the saturated zone, which is a zone where all voids are filled with water.

H

hydroelectric power water use The use of water in the generation of electricity at plants where the turbine generators are driven by falling water.

hydrologic cataloging unit An eight-digit cataloging unit that identifies a geographic area representing part or all of a surface drainage basin, a combination of basins, or a distinct hydrologic feature. Sometimes known as a watershed.

I

industrial water use Water used for industrial purposes such as fabrication, processing, washing, and cooling, and includes such industries as steel, chemical and allied products, paper and allied products, smelting, and petroleum refining. The water may be obtained from a public supply or may be self supplied. *See also* public-supply water use and self-supplied water.

instream use Water that is used within a stream channel for such purposes as hydroelectric power generation, navigation, water-quality improvement, fish propagation, and recreation. Sometimes called nonwithdrawal use or in-channel use.

irrigation district A cooperative, self-governing public corporation with definite geographic boundaries and taxing power. Its function is to obtain and distribute water for irrigation of lands within the district.

irrigation system Equipment used to distribute water to crops or other irrigated lands. Irrigation systems are grouped into the following three broad categories:

> **sprinkler** An irrigation system in which water is applied by means of perforated pipes or nozzles operated under pressure so as to form a spray pattern.

> **surface** Irrigation by means of flood, furrow, or gravity. *Flood* irrigation is the application of irrigation water in which the entire soil surface is covered by ponded water. *Furrow* is a partial surface-flooding method of irrigation in which water is applied in furrows or rows of sufficient capacity to contain the design irrigation stream. *Gravity* is an irrigation method in which water is not pumped, but flows in ditches or pipes and is distributed by gravity.

> **microirrigation** An irrigation system that wets only a discrete portion of the soil surface in the vicinity of the plant by means of applicators operated under low pressure. The applicators can be placed on or below the surface of the ground or can be suspended from supports. Subsurface systems that control the height of the water table are included in this category.

irrigation water use Application of water on lands to assist in the growing of crops and pastures or to maintain vegetative growth on recreational lands such as parks and golf courses. Includes water applied for pre-irrigation, frost protection, chemical application, leaching salts from the root zone, and dust suppression, as well as water lost in conveyance. Also includes irrigation for cemeteries, turf farms, and other landscaped areas but does not include domestic lawns and gardens, which are included in the domestic water-use category.

K

kilowatt-hour (kWh) A unit of energy equivalent to 1,000 watt-hours.

L

livestock water use Water for livestock watering, feedlots, dairy operations, and other on-farm needs. Livestock includes cattle, sheep, goats, hogs, poultry, horses, and fur-bearing animals. *See also* animal specialties water use.

M

makeup water The water pumped into a closed-loop cooling system to replace the circulating water lost by evaporation, drift, blowdown, and leakage. Makeup water usually is expressed as a percentage of the total amount of water circulated.

mining water use Water used for the extraction of naturally occurring minerals including solids, such as coal and ores; liquids, such as crude petroleum; and gases, such as natural gas. Also includes uses associated with quarrying, well operations, milling, and other preparations customarily done at the mine site or as part of mining activity. Mining water use does not include water used in processing, such as smelting, refining petroleum, or slurry pipeline operations, which are included in industrial water use.

N

North American Industry Classification System (NAICS) codes Hierarchical codes established in 1997 by the Office of Management and Budget in cooperation with its counterparts in Canada and Mexico. NAICS are used in the classification of establishments by type of activity in which they are engaged, thus enabling comparison of industries from the three countries. NAICS replaces the Standard Industrial Classification (SIC) system. *See also* Standard Industrial Classification (SIC) codes.

O

offstream use Water withdrawn or diverted from a surface-water source for public water supply, domestic, industry, irrigation, livestock, thermoelectric power generation, and other uses. Sometimes called off-channel use or withdrawal.

once-through cooling system A cooling system in which water is withdrawn, circulated through heat exchangers, then returned to a body of water at a higher temperature. Also referred to as an open-loop cooling system.

P

per-capita water use The average amount of water used per person during a standard time period, generally per day. Per-capita use may be calculated on the basis of total water use, public-supply water use, self-supplied domestic water use, or domestic deliveries from public supply.

pre-irrigation The application of water to cropland before planting to assure adequate soil moisture for crop germination and early plant growth.

public-supply water use Water withdrawn by public and private water suppliers that furnish water to at least 25 people or have a minimum of 15 connections. Public suppliers provide water for a variety of uses, such as domestic, commercial, industrial, and thermoelectric power, public use, and losses. See also commercial water use, domestic water use, industrial water use, thermoelectric power water use, and public water use.

public water use Water provided by a public supply for such uses as firefighting, street washing, water treatment, municipal buildings, parks, and swimming pools. Generally, public water use is not billed by the public water supplier. *See also* public-supply water use.

R

reclaimed wastewater Wastewater-treatment plant effluent that has been diverted for beneficial use before it reaches a natural waterway or aquifer.

recycled water Water that is used more than once after withdrawal and before it returns to the natural hydrologic system.

residential water use *See* domestic water use.

return flow Water that reaches a ground-water or surface-water source after it is released from the point of use, and thus becomes available for further use.

reuse Use of water that has undergone wastewater treatment and is delivered to a user as reclaimed wastewater. *See also* reclaimed wastewater and recycled water.

S

saline water Water that contains 1,000 mg/L or more dissolved solids. *See also* freshwater.

self-supplied water Water that is withdrawn directly from a ground-water or a surface-water source by a user, as opposed to water that is delivered by a public supplier.

Standard Industrial Classification (SIC) codes
Four-digit codes established by the Office of
Management and Budget, last revised in 1987,
and used in the classification of establishments
by type of activity in which they are engaged.
SIC codes are being replaced by NAICS codes.

surface water An open body of water, such as
a stream, lake, or reservoir.

T

thermoelectric power water use Water used
in the process of generating electricity with
steam-driven turbine generators. The water
may be obtained from a public supply or may
be self supplied. See also public-supply water
use and self-supplied water.

transpiration Process by which water that is
absorbed by plants, usually through the roots,
is evaporated into the atmosphere from the
plant surface. *See also* evaporation and evapotranspiration.

U

unaccounted-for water The difference
between the amount of water that a public-
supply facility withdraws and the amount of
water that is accounted for by metered uses.

Also known as public-supply residual or public
use and losses.

W

wastewater treatment Removal or reduction
of solids, pathogens, or other undesirable con-
stituents from wastewater.

water transfer Conveyance of water from one
area to another using natural or manmade
channels.

water treatment Processes such as filtration
and disinfection of water prior to delivery and
use.

water-use coefficient A factor or ratio used to
estimate a quantity of water used on the basis
of a related quantity. Examples of water-use
coefficients include daily per-capita water use,
crop water needs, livestock water requirements, water use per employee, and water use
per unit of product.

withdrawal The removal of ground water or
surface water from the natural hydrologic
system for uses including public supply, commercial, domestic, industry, irrigation, mining,
livestock, aquaculture, and thermoelectric
power generation. See also offstream use.

Guidelines for Preparation of State Water-Use Estimates for 2000

Edited by Joan F. Kenny

Abstract

This report describes the water-use categories and data elements required for the 2000 national water-use compilation conducted by the U.S. Geological Survey (USGS) as part of its National Water Use Information Program. It identifies sources of water-use information, guidelines for estimating water use, and required documentation for preparation of the national compilation by State for the United States, the District of Columbia, Puerto Rico, and the U.S. Virgin Islands. The data are published in USGS Circular 1268, *Estimated Use of Water in the United States in 2000*. USGS has published circulars on estimated use of water in the United States at 5-year intervals since 1950.

As part of this USGS program to document water use on a national scale for the year 2000, all States prepare estimates of water withdrawals for public supply, industrial, irrigation, and thermoelectric power generation water uses at the county level. All States prepare estimates of domestifc use and population served by public supply at least at the State level. All States provide estimates of irrigated acres by irrigation system type (sprinkler, surface, or microirrigation) at the county level. County-level estimates of withdrawals for mining, livestock, and aquaculture uses are compiled by selected States that comprised the largest percentage of national use in 1995 for these categories, and are optional for other States. Ground-water withdrawals for public-supply, industrial, and irrigation use are aggregated by principal aquifer or aquifer system, as identified by the USGS Office of Ground Water.

Some categories and data elements that were mandatory in previous compilations are optional for the 2000 compilation, in response to budget considerations at the State level. Optional categories are commercial, hydroelectric, and wastewater treatment. Estimation of deliveries from public supply to domestic, commercial, industrial, and thermoelectric uses, consumptive use for any category, and irrigation conveyance loss are optional data elements. Aggregation of data by the eight-digit hydrologic cataloging unit is optional.

Water-use data compiled by the States are stored in the USGS Aggregated Water-Use Data System (AWUDS). This database is designed to store both mandatory and optional data elements. AWUDS contains several routines that can be used for quality assurance and quality control of the data, and also
produces tables of water-use data compiled for 1985, 1990, 1995, and 2000. These water-use data are used by USGS, other agencies, organizations, academic institutions, and the public for research, water-management decisions, trend analysis, and forecasting.

Introduction

The U.S. Geological Survey (USGS) has compiled and published estimates of water use for the Nation at 5-year intervals since 1950. In 1977, the Congress provided funding for USGS to establish the National Water Use Information Program, which is a cooperative effort with States to collect reliable and uniform water-use information. Data collected for each State in the United States, the District of Columbia, Puerto Rico, and the U.S. Virgin Islands are presented in USGS Circulars entitled, *Estimated Use of Water in the United States* (MacKichan, 1951, 1957; MacKichan and Kammerer, 1961; Murray, 1968; Murray and Reeves, 1972, 1977; Solley and others 1983, 1988, 1993, 1998; and Hutson and others, 2004). The circular for 2000 marks the completion of a 50-year span of national water-use reports in this series.

Estimates of ground-water and surface-water withdrawals are mandatory for all States for the public-supply, domestic, industrial, irrigation, and thermoelectric power categories. Ground-water withdrawal estimates from principal aquifers or aquifer systems (U.S. Geological Survey, 2003) are mandatory for the public-supply, industrial, and irrigation categories. Estimates of ground-water and surface-water withdrawals for the mining, livestock, and aquaculture categories are mandatory for selected States and optional for all other States. The commercial, hydroelectric power, and wastewater treatment categories are optional for all States and are not included in the circular for 2000. This report describes guidelines for estimating data for the categories that are mandatory for all or selected States.

In some States, water-use information for many categories is collected by government agencies, universities, or private organizations. For other categories, water-use information may be obtained from multiple sources or estimated using coefficients and ancillary data. Water-use project chiefs in each USGS State office are familiar with the availability and reliability of information in their own States and work with their cooperators to produce reliable estimates of water use. This information may be electronically stored in databases developed by individual States or by the USGS.

The Aggregate Water-Use Data System (AWUDS) is a USGS database that is specifically designed to store and manipulate the aggregate water-use information compiled for the 5-year reports. AWUDS also is capable of storing annual aggregate water-use data. During previous compilations, estimates of water use were aggregated by county and hydrologic cataloging unit within each State, and published by State and water-resources region. For the 2000 compilation, water-use estimates that are aggregated by county or State are published by State. Estimates of ground-water use that are aggregated by principal aquifer or aquifer system are published in a separate report.

Purpose and Scope

The purpose of this report is to provide guidelines for preparation of State water-use estimates. Water-use categories, data elements, aggregation levels, and documentation requirements for the 2000 national water-use compilation are defined. This information is useful both to those who prepare the estimates and to those who use the data. The scope of the 2000 compilation was modified from previous years in response to budget considerations and concentrates on withdrawals for the public-supply, domestic, industrial, irrigation, and thermoelectric power water-use categories in each State. Withdrawals for the livestock, aquaculture, and mining water-use categories are mandatory for selected States comprising at least 75 percent of those category totals for 1995, and fewer data elements and aggregation levels are compiled for some categories. Withdrawal estimates are aggregated at the county, State, or aquifer level. Aggregation by hydrologic cataloging unit (described in Seaber and others, 1987) is optional. Estimates of consumptive use, irrigation conveyance loss, reclaimed wastewater, and deliveries from public water suppliers also are optional for the 2000 compilation. Efforts have been made to improve quality assurance/quality control of the estimates and to ensure adequate documentation of data sources and compilation methods.

This report also summarizes guidance provided to USGS State water-use project chiefs in the effort to compile nationally consistent water-use estimates for the report *Estimated Use of Water in the United States in 2000* (Hutson and others, 2004), hereinafter referred to as USGS Circular 1268. This guidance was prepared by regional water-use specialists, with support from the USGS Office of Ground Water, and was presented to personnel in each USGS State office at the start of the compilation effort to describe consistent procedures for completing the 2000 estimates. The methods presented were general in nature and were not intended to replace more specific or detailed methodologies that may be used by some project chiefs. Because data availability and reliability vary from State to State, each project chief documented the specific sources of data and compilation methods. These unpublished documents are maintained by each State as a reference for current data and as a starting point for continued data collection.

Acknowledgments

This report was edited by Joan F. Kenny from material developed over time by personnel from the USGS National Water Use Information Program. Contributors to this text include Nancy L. Barber, E. James Crompton, Terrance W. Holland, Susan S. Hutson, Joan F. Kenny, Kristin S. Linsey, Deborah S. Lumia, Molly A. Maupin, Wayne B. Solley, and William E. Templin.

Water-Use Compilation Requirements for 2000

Water-Use Categories and Data Elements

Water-use categories for which data may be collected for the 2000 compilation and the data elements included in each category are summarized in table 1 in the "Supplemental Information" section at the back of this report. The national dataset presented in USGS Circular 1268 consists of the mandatory data elements for the public-supply, domestic, industrial, irrigation, thermoelectric power, mining, livestock, and aquaculture water-use categories. State data sets may include any optional data elements and categories for which the State project chiefs compiled information.

Ground- and surface-water withdrawal estimates for the public-supply, industrial, irrigation, and thermoelectric power categories are mandatory at the county level for all States. Estimates of fresh ground-water withdrawals for the public-supply, irrigation, and industrial categories also are compiled by principal aquifer or aquifer system in all States.

Total population served by public water suppliers is mandatory at the State level only. States optionally may compile individual county estimates for population served by public supply, as well as separate populations served by ground and surface water.

Withdrawal estimates for domestic use are mandatory at the State level only. Self-supplied domestic population is calculated as the difference between total State population and population served by public suppliers. States optionally may compile county estimates for domestic withdrawals. If the population served is estimated by county, then self-supplied population is calculated by county. Aggregation of domestic ground-water withdrawals by principal aquifer also is optional.

Mandatory water-use categories for selected States are mining, livestock, and aquaculture. Withdrawal estimates for the mining, livestock, and aquaculture categories are mandatory at the county level for those States comprising more than 75 percent of the 1995 national total for those uses. Data for these three categories optionally may be compiled for other

States but are included in the national data set only if collected for the entire State. Estimates of ground-water withdrawals by principal aquifer also are optional for these categories.

Compilation of saline withdrawals, defined as water with dissolved solids of 1,000 mg/L or more, is mandatory for the industrial, thermoelectric power, and mining categories in States where saline water is used. Saline withdrawals for public supply are included with estimates of freshwater withdrawals or optionally may be stored separately in AWUDS.

The commercial, hydroelectric power, and wastewater treatment categories are optional for the 2000 compilation, although estimates for these categories may be compiled and stored in the AWUDS database. Estimates of reservoir evaporation, a category included in previous compilations, are not made for 2000, and this category cannot be stored in AWUDS.

Estimates of consumptive use, irrigation conveyance loss, reclaimed wastewater, and deliveries from public suppliers are optional for the 2000 compilation. Although these data elements are important, insufficient information is available in many cases to make these estimates consistently in all States. Estimates aggregated by eight-digit hydrologic cataloging unit are optional. These minimum requirements for the 2000 national compilation are designed to allow for a more cost-effective and nationally consistent product from individual States. The modified approach is not intended to limit any efforts to meet the needs of local programs and cooperators.

Units of Measurement

Water withdrawals are reported in million gallons per day (Mgal/d). Population is reported in thousands of people. Acres irrigated are reported in thousand acres. All values are reported to two places after the decimal point. A value of zero usage for a data element indicates either no usage, usage of less than 0.01 Mgal/d, or that no data are collected.

Aggregation Levels

Withdrawals for the public-supply, irrigation, industrial, thermoelectric power, mining, livestock, and aquaculture categories are compiled at the county level for each State. For this compilation, the District of Columbia, Puerto Rico, and the Virgin Islands are treated as States, and Louisiana parishes and Alaska boroughs are treated as counties. Domestic withdrawals and population served by public supply are mandatory at least at the State level but may be compiled at the county level. All data are summarized and published in USGS Circular 1268 at the State level.

Estimates of ground-water withdrawals of freshwater for public supply, industrial, and irrigation uses are aggregated for approximately 68 principal aquifers and aquifer systems, listed in table 2 (in the "Supplemental Information" section), that are identified in the National Atlas (U.S. Geological Survey, 2003). A designation of "other" is used to aggregate any withdrawals

that cannot be assigned to one of these principal aquifers or aquifer systems.

AWUDS Database

Aggregate water-use data have been stored electronically by the USGS since 1980. The AWUDS database was created in 1985 to store data aggregated by county or hydrologic cataloging unit. AWUDS was redesigned for 2000 and released as a personal computer-based system installed in each USGS State office. State water-use project chiefs are required to enter aggregate data by county or State for all mandatory water-use categories into the AWUDS database. Additionally, AWUDS is able to store any optional data aggregated by county, aquifer, or eight-digit hydrologic cataloging unit that the State water-use project chiefs may collect.

Water-use data for years other than the compilation years may be stored in the redesigned AWUDS database, which allows greater utility of the software for individual State water-use programs. Data may be entered and edited interactively or imported from external files. AWUDS can generate various reports to table the data by category or area, show entered data elements, and provide calculated values for certain categories. Quality-assurance programs within AWUDS include checks for erroneous values, comparison of totals by area, and comparison of data between 2 years.

Documentation

Documentation of data sources and compilation methods for each State is required as part of the 2000 compilation. Documentation of how the data are compiled serves to substantiate the data published in USGS Circular 1268 and provides a valuable record for subsequent compilations.

The format for each State's documentation is determined by the water-use project chief. All documentation must include the following elements:

- Sources of data and coefficients used;
- Agency and contact information;
- Methods of compiling and estimating data; and
- Location and types of data files.

State water-use project chiefs are required to transmit completed documentation in electronic form to the respective USGS regional water-use specialist. If the 2000 water-use estimates differ substantially from those reported in 1995, possible explanations are to be identified in this documentation.

Compilation Methods

General Techniques

For many categories, water-use data can be collected by either a site-specific inventory or a representative survey. The most accurate and desirable compilation technique is a complete site-specific inventory, which can be aggregated to the desired level. In many States, site-specific data are unavailable, incomplete, or only available aggregated by political units. For some categories, site-specific data are available only for the largest users; this partial site-specific inventory may account for much of the water used in a particular category. A representative survey is a selective sampling of typical water users in a water-use category. Representative site-specific data are collected and used to develop water-use coefficients (for example, water used per person during a specified period of time) related to known ancillary information such as population, number of employees, crop acreages, or livestock counts. For example, a coefficient of water use per person in a specified period of time can be used to estimate domestic water use by a known population. Water use can be estimated in areas that lack site-specific data by using coefficients developed in areas with good data. This method provides acceptable results if the water-use and ancillary data characteristics are similar for the different areas.

In some cases, no site-specific data are available for a category. Water withdrawals may be estimated entirely on the basis of ancillary data and water-use coefficients. This technique is acceptable if the ancillary data are uniformly available for a State and the coefficients are documented.

The 2000 national compilation requires assimilation of a large amount of information. To produce the most complete and defensible data for each State, the greatest effort and time are spent collecting information about the largest users and the largest categories of use in individual States. Compilation of accurate data for the largest public suppliers, industries, agricultural regions, and powerplants produces State totals that are fairly reliable. As time and resources permit, data on smaller users within each category are collected.

A glossary of terms used in this report and in preparing estimates of water use is shown at the front of this report. The glossary is not all-inclusive but focuses on terms commonly used to explain or describe water-use characteristics. Some terms that are defined but not used in this report represent usage from previous compilations.

Internet Resources

Numerous agencies and organizations are listed in this report as possible sources of information for compiling water-related data for the various categories of use. Many of these agencies and organizations maintain information and databases on the Internet. Specific Internet addresses are listed in this report as they appeared on the date last accessed.

Withdrawals by Aquifer

Fresh ground-water withdrawals by principal aquifer or aquifer system for the categories of public supply, industry, and irrigation are mandatory for the 2000 national water-use compilation. Estimates of saline ground-water withdrawals are not aggregated by principal aquifer or aquifer system. Generally, total ground-water withdrawals by category are estimated first and then allocated among the principal aquifers or aquifer systems identified for each State. In the absence of site-specific data, the preferred approach is to: (1) determine fresh ground-water withdrawals for these three categories, usually at the county level, using the best available data, (2) integrate information about aquifer properties (areal extent, water quality, and depth to water) with information about the water-use categories, and (3) allocate withdrawals to the appropriate aquifers. Some methods developed for assigning withdrawals to principal aquifers, as well as evaluating their accuracy and reliability, are described in the "Aquifer Withdrawals" section of this report. Estimates of aquifer withdrawals are reviewed for reasonability by USGS staff in each State office.

Quality Assurance/Quality Control

Quality assurance/quality control (QA/QC) is an important step in the compilation of water-use estimates. Errors can occur in both the compilation of the data and also in data entry. The following basic checks can be made during data compilation and after AWUDS data entry.

- Use AWUDS quality-assurance programs to check for erroneous data and to compare State totals with aquifer totals by category.

- Review spatial distribution of data using choropleth or other types of geographic information system (GIS) maps.

- Compare 2000 data with those from the previous compilations. Examine changes on the basis of percentage change, statistical analysis, or general patterns within a State.

- Use sorting routines to check for possible errors in largest and smallest withdrawals.

- Examine calculated values such as per-capita use and irrigation application rates.

- Check county and State data for population served to ensure that they do not exceed total census populations.

Industrial Classification Coding Systems

The Standard Industrial Classification (SIC) coding system was developed to promote the comparability of establishment data describing various facets of the United States economy (Office of Management and Budget, 1987). The SIC system was intended to cover the entire field of economic activities—agriculture, forestry, fishing, hunting, and trapping; mining; construction; manufacturing; transportation, communications, electric, gas, and sanitary services; wholesale trade; retail trade; finance, insurance, and real estate; personal, business, professional, repair, recreation, and other services; and public administration. The SIC system is based on the primary activity in which the industry is engaged. The structure of the classification makes it possible to tabulate, analyze, and publish industry data on the basis of a two-digit major group, a three-digit industry group, or a four-digit industry code.

The USGS has assigned each four-digit industry code in the 1987 SIC manual (Office of Management and Budget, 1987) to a water-use category. In some cases, a code may be listed under more than one water-use category. The SIC codes can be useful for assigning withdrawals to the public supply, industrial, mining, and thermoelectric power categories. A list of SIC codes by water-use category is shown in table 3 in the "Supplemental Information" section at the back of this report.

A newer system of classification, the North American Industry Classification System (NAICS), went into effect in 1997 in the United States. Descriptions of the NAICS codes and correlation with SIC codes are published by the U.S. Census Bureau (2000, 2002). However, USGS water-use project chiefs may continue to identify industries using SIC codes for the 2000 compilation.

Public Supply

Category Description

Public supply refers to water withdrawn from ground and surface sources by public and private water systems for use by cities, towns, rural water districts, mobile-home parks, Indian reservations, and military bases. Public-supply facilities provide water to at least 25 people or have a minimum of 15 service connections, and are classified as SIC 4941 (table 3). Water withdrawn by public suppliers may be delivered to users for domestic, commercial, industrial, and thermoelectric purposes, as well as to other public water suppliers. Some public supply water is used for water treatment, wastewater treatment, public services such as pools, parks, and city buildings, or is lost through system leaks, unmetered services, and maintenance.

Data Elements

Mandatory

- Ground-water withdrawals, freshwater, by county
- Surface-water withdrawals, freshwater, by county
- Ground-water withdrawals, freshwater, by aquifer
- Total population served, by State

Optional

- Population served, by county
- Population served by ground water, by county
- Population served by surface water, by county
- Ground-water withdrawals, saline water, by county
- Surface-water withdrawals, saline water, by county
- Number of facilities, by county
- Reclaimed wastewater, by county
- Any data aggregated by hydrologic cataloging unit

Sources of Information

Information on public-supply withdrawals, sources of water, and population served may be obtained from a variety of sources. The following lists some of the most commonly used sources of data for the compilation. This list is not all inclusive, and USGS water-use project chiefs are encouraged to investigate any other potential sources of information available for their States.

- Individual public water suppliers
- State agencies that administer water rights, allocate water to users, or collect water-use data
- State agencies that enforce the Safe Drinking Water Act and issue permits for water discharge
- State agencies that regulate utility rates
- County planning and zoning agencies
- Community planning and development agencies
- U.S. Census Bureau: <http://www.census.gov/>

  American FactFinder database, the U.S. Census Bureau's online data source, containing information on population and housing: <http://factfinder.census.gov/home/saff/main.html?_lang=en>

- U.S. Environmental Protection Agency (USEPA): <http://www.epa.gov/>

  Safe Drinking Water Information System (SDWIS):

    Access to data that States must report to USEPA as required by the Safe Drinking Water Act.

    Safe Drinking Water Query Form: <http://www.epa.gov/enviro/html/sdwis/sdwis_query.html>
    Search the SDWIS database for selected information. Includes data on water system name, county, population served, and primary water source type. Includes community/noncommunity and transient/nontransient water systems.

  Office of Ground Water and Drinking Water: <http://www.epa.gov/OGWDW/>

Water Discharge Permits, Permit Compliance System (PCS) database:

<http://www.epa.gov/enviro/html/pcs/pcs_query_java.html>

Access to data regarding facilities holding National Pollutant Discharge Elimination System (NPDES) permits. Specify the facilities by using any combination of facility name, geographic location, SIC code, and chemicals.

When a list of water systems is retrieved for a specific State, the State agency responsible for regulating drinking water will be stated at the top of the list.

- American Water Works Association:
  <http://www.awwa.org/>

  Links to local Web sites and contacts:
  <http://www.awwa.org/sections/sechome.htm>

  Links to information on individual public suppliers:
  <http://www.awwa.org/community/links.cfm>

- Association of State Drinking Water Administrators (ASDWA):

  <http://www.asdwa.org/>

  Links to a variety of Web resources including a page with links for each State's drinking water program pages and the primary State agency responsible for drinking water.

- National Rural Water Association:
  <http://www.nrwa.org/>

- Organizations whose membership rosters include public or private utilities, such as League of Municipalities or State Rural Water Associations.

Compilation Techniques

Required data elements for the public-supply category include withdrawals from ground and surface sources and population served. An important first step in compiling these data for particular States is to develop a list of the names and locations of public water suppliers, populations served, and sources of water. These data may be available from one or more State agencies, from previous compilations, or from USGS water-use databases. USEPA maintains the SDWIS database using data obtained from State agencies that administer water-quality regulations for drinking water. The SDWIS database includes information on the source of supply, location of water intakes, and retail population served for all community water systems (CWS), which are those serving at least 25 people or 15 or more service connections on a year-round basis. SDWIS does not include any information on quantity of water withdrawn. In some cases, the State agencies providing the data to USEPA may have more up-to-date and complete information, including withdrawals.

Withdrawals

Public-supply withdrawals are compiled for the county in which the withdrawals occurred, although the water may have been subsequently distributed and used elsewhere. Individual public water suppliers usually can provide the most accurate information on sources of water and amounts withdrawn. Cooperator surveys or reporting programs are a good source for this information in States where water suppliers are required to report usage. Individual State agencies have varying criteria for obtaining water-use data; these criteria may be based on permit requirements, size of system, or magnitude of withdrawals. Return rates on cooperator surveys, as well as the amount of quality assurance the surveys receive, also vary from State to State. For these reasons, some information may need to be obtained directly from suppliers or estimated.

USGS water-use project chiefs may survey public water suppliers by telephone calls or visits. Mail surveys may be used if a State cooperator participated in the mailing, or if requirements of the 1995 Paper Reduction Act (U.S. Congress, 1995) are met. Usually, information can be obtained from a director of public works, city engineer, city clerk, bookkeeper, manager, or operator. The level of detail obtained depends on the amount of time available to collect public-supply data and the ability of the people contacted to provide this information. Useful information to request from these contacts includes the following list of items.

- **Source(s) of water.** Well information (local name or number, depth, location) and names of aquifers used for ground-water sources, names and locations of surface-water sources (streams, lakes, reservoirs), and sources of any purchased water. Some water suppliers use combinations of ground water, surface water, and (or) purchased water.

- **Total withdrawals in 2000.** Amounts of water pumped may have been measured by flowmeters or calculated using pump rates and number of hours pumped. Data may be available by day, month, or year.

- **Metering points.** Water may be metered before treatment, during transport to another location, or as it is delivered to users. It is important to know the metering point to avoid exclusion of treatment or transit losses, or double counting of water.

- **Type of treatment plant.** Surface water usually requires more water for treatment processes, such as backwashing filters, than does ground water. If total reported water produced represents finished rather than raw water, an amount or percentage of water for treatment use may need to be added for systems with large water-treatment usage.

- **Names of other water suppliers and amounts of water transferred.** Many water suppliers purchase and sell water.

- **Numbers of service connections of various types.** Numbers of active residential meters, including single-family and multi-family housing, are useful for determining population served. Numbers of commercial and industrial connections are useful for determining component uses that affect the magnitude of withdrawals.

- **Amounts of water sold in 2000 to various end uses.** Deliveries include residential, commercial, industrial, and free or public uses. If delivery information is available, it can be used to develop water-use coefficients for estimating withdrawals from similar, nonsurveyed suppliers. Data on total metered uses also can be subtracted from total withdrawals to determine amounts of unaccounted for water.

- **Locations of retail service areas.** Many public water suppliers serve customers outside city limits or in multiple counties or States.

The largest water suppliers should be contacted to obtain data or to verify data obtained from State agencies or other databases. Any estimation of withdrawals for large suppliers should consider the variable amounts of water that may be sold to other suppliers, delivered to residential, commercial, industrial, or thermoelectric power users, used for public uses, and lost. For smaller public water suppliers that are not surveyed, water use may be estimated on the basis of water allocations, average production values, previously reported usage, or population changes. If estimates are based on residential use in nonsurveyed systems, estimates of other types of deliveries, public use, and system losses should be added to these estimates to approximate total water use. Information on commercial and industrial activity may be derived from local knowledge, State directories of business and manufacturers, or purchased databases.

The difference between total raw water and metered uses often is referred to as "unaccounted for" water, a term that is equivalent to "public use and losses" in USGS Circular 1200 for 1995 water use (Solley and others, 1998). This publication indicates that public use and losses averaged 15 percent of public-supply withdrawals in the United States in 1995 (Solley and others, 1998). Public use and losses vary substantially among water suppliers, depending on treatment needs, system efficiency, and the amount of public water use. In general, systems that withdraw surface water have greater losses due to treatment than do systems that withdraw ground water. Older systems and those that are undergoing repairs to lines or towers have greater losses due to leaks, flushing, and tower draining than systems with few problems. Public uses include water, often used free of charge, for public buildings, fire fighting, irrigation of city parks, golf courses and ball fields, municipal pools, and water treatment and wastewater treatment. Unless a water sup-

plier meters its free uses, public use and system losses cannot be identified separately.

Information on water transfers is very important for estimating public-supply withdrawals. Estimates of total water use for suppliers using a combination of their own withdrawals and purchased water, or for those selling water wholesale, need to be adjusted to identify withdrawals by county in which they occur.

Per-Capita Use Coefficients

Public-supply withdrawals often are estimated using water-use coefficients such as average per-capita use or average use per meter. These coefficients are used with current information on population or number of meters to estimate water use. An average per-capita use coefficient (gallons per person per day) for a given system or area is determined by dividing the gallons of water withdrawn during a specified time period by the population served and the number of days. Per-capita use coefficients that are based on total public-supply withdrawals are larger than coefficients that are based only on residential use because the total includes all other deliveries and losses. Per-capita use coefficients are generally larger for systems that serve large industrial or commercial users or have large losses.

An overall per-capita use coefficient for a system, determined simply as total withdrawals divided by population served, is not meaningful if there are any water transfers. Average county per-capita use coefficients (total withdrawals in a county divided by total population served by public suppliers in that county) are not representative if any water is exported for use in a different county than the one in which it was withdrawn, or if any population is served by water withdrawn in another county.

Residential per-capita use coefficients can be determined using domestic delivery data from surveyed water suppliers. If possible, it is advantageous to develop these coefficients using information from water suppliers of representative sizes and geographic locations because per-capita use often is affected strongly by climate, water rates, and level of customer affluence. Residential per-capita use coefficients also may be available from previous compilations in each State or from other published material. Some States have published projections of future water demands and may have calculated per-capita use rates. Regional and State planning agencies, State natural resource agencies, consulting firms, or the State's public health agency are other possible sources for residential per-capita use coefficients.

Outdoor water use is a significant component of residential per-capita usage, especially in arid climates. In some areas, residential communities may receive potable water from a public water supplier for indoor use and nonpotable water through a separate distribution system for outdoor irrigation of lawns and gardens. Typically, developers provide the nonpotable water to houses in urban areas built on former farmland, using water previously allocated for agricultural irrigation use. Public-supply per-capita use coefficients in these "dual-use" areas generally are lower than in areas where customers use publicly supplied water for both indoor and outdoor irrigation uses.

Population Served

Populations served by public suppliers are estimated for all public water suppliers regardless of the source of their water (ground water, surface water, purchased water, or a combination). Population served refers to the resident population receiving water at retail from a public supplier on a year-round basis and excludes vacationers and second-home owners. People living on military bases, on Indian reservations, and in prisons constitute a population served by the supplier serving these communities. Population served is compiled in the county of residence, which is not necessarily where the withdrawals occur.

For the 2000 compilation, population served is required to be reported at the State level. However, a valid State total depends on collection of information for individual water suppliers. Information on populations served may be obtained from the U.S Census Bureau, individual public water suppliers, State cooperators, or SDWIS. Reported populations from each of these sources need to be checked for reliability. Service areas for many public suppliers do not correspond to political boundaries such as city limits or county lines; therefore, census populations need to be adjusted to accommodate customers living outside city or county lines to account for total population served. For water suppliers with service areas that span more than one county, an estimate is made of the population residing in each county.

Populations served also can be estimated on the basis of numbers of occupied residential connections. This technique is useful for determining population in service areas for which no precise population is available, such as large water suppliers that serve parts of multiple metropolitan areas, rural water districts, and self-supplied housing developments. Individual water suppliers can provide numbers of residential service connections, which may include multi-family dwellings. Census data on housing characteristics for incorporated places include ratios of average persons per occupied residential connection. Populations reported by public water suppliers may be over reported if they are estimated using ratios of people per household that are too large. Other errors may be caused by double counting people served by more than one system or by wholesale supplies. A population figure may have been reported by each of several suppliers whose service areas overlap. Populations reported by a supplier providing wholesale supplies to other systems may erroneously include both retail and wholesale populations and need to be adjusted.

Errors in reported populations served can be detected when comparing aggregated county population served to total county census populations. This comparison is needed for quality assurance of the State total. County population served can-

not exceed the total county population. State population served cannot exceed the total State population.

Domestic

Category Description

Domestic water use is water used for both indoor and outdoor household purposes. Common indoor uses include drinking, food preparation, bathing, washing clothes and dishes, and flushing toilets. Major outdoor uses include watering lawns and gardens and washing cars. Water for domestic water use may be delivered from a public supplier or self supplied if obtained from a private source such as a well.

Data Elements

Mandatory

- Ground-water withdrawals, freshwater, by State
- Surface-water withdrawals, freshwater, by State

Optional

- Ground-water withdrawals, freshwater, by county
- Surface-water withdrawals, freshwater, by county
- Deliveries from public supply, by county
- Consumptive use, by county
- Any data aggregated by hydrologic cataloging unit

Sources of Information

Information concerning domestic water use may be obtained from the sources in the following list. This list is not all inclusive, and water-use project chiefs are encouraged to investigate any other potential sources of information available for their States.

- Well records of individual households from cooperator permits
- Water supplier records of residential use
- Water supplier records of population served
- State pollution control agencies
- State agency responsible for water permitting
- Ground-water resources studies
- Local chamber of commerce
- State public health agencies
- Sewage treatment facilities
- Tax appraiser data

- Planning agencies
- U.S. Census Bureau:
  <*http://www.census.gov/*>

  American FactFinder, the U.S. Census Bureau's online data source:
  <*http://factfinder.census.gov/home/saff/ main.html?_lang=en*>

  Summary File 1, in "Data Sets," includes selected population and housing characteristics to the block/census tract level.
- American Water Works Association:
  <*http://www.awwa.org/*>

  Water-use coefficients:
  <*http://www.awwa.org/Advocacy/pressroom/ index.cfm*>

  WaterWiser, the Water Efficiency Clearinghouse:
  <*http://www.waterwiser.org/*>
- U.S. Environmental Protection Agency (USEPA):
  <*http://www.epa.gov/*>

  Water Discharge Permits, Permit Compliance System (PCS) database:
  <*http://www.epa.gov/enviro/html/pcs/ pcs_query_java.html*>

  Access to data regarding facilities holding National Pollutant Discharge Elimination System (NPDES) permits. Specify the facilities by using any combination of facility name, geographic location, SIC code, and chemicals.

Compilation Techniques

Self-supplied domestic water withdrawals typically are estimated by multiplying the self-supplied population by a per-capita use coefficient. For the 2000 compilation, the State self-supplied population is calculated by subtracting the total population served by public supply from the total census population for the State. This residual population then can be multiplied by a statewide per-capita use coefficient. Use of county figures for population served and total population allows an estimate of county self-supplied population that can be used with different per-capita use coefficients for various geographic regions within a State.

Reliable figures for population served by public supply yield reliable figures for self-supplied population. Vacationers and users of second homes in publicly supplied areas generally are not included in the population served because they already have been counted in the location of their primary residence. The estimated self-supplied population may be too small if the population served by public suppliers includes nonresidents. Local planners or consulting firms may be able to provide an estimate of the percentage of second homes in areas with substantial numbers of nonresidents.

Self-supplied domestic withdrawal estimates may be more accurate if per-capita use is determined for different geographic areas of the State that have different climates or water-use characteristics. Per-capita use is greater in more arid regions where outdoor watering is a large component of household water use. A preferred method for developing domestic per-capita use coefficients is to obtain data on domestic deliveries and population served from different water suppliers throughout a State. Domestic deliveries divided by the population served will give an estimate of per-capita use. Per-capita use coefficients that are based on reported delivery data may differ from self-supplied per-capita use depending on the degree of outdoor watering or may be affected by ordinances controlling outdoor use. This method is more time consuming than using a uniform State coefficient but yields per-capita use coefficients that reflect climatic conditions and local patterns of domestic use. If per-capita use rates are similar throughout the State, an average per-capita use for the entire State may be appropriate. If per-capita use rates indicate regional differences within the State, an average per-capita use can be estimated for each county or region.

Domestic per-capita use coefficients also may be obtained from estimates provided by other agencies or available in the literature. Some States have published projections of future water demands and may have calculated per-capita use rates. Regional and State planning agencies, State natural resource agencies, or the State's public health agency are possible sources for this information. Per-capita use coefficients also have been determined as part of research conducted by professional organizations such as the American Water Works Association and by consulting firms.

Self-supplied domestic withdrawals sometimes can be estimated using information from wastewater-treatment facilities that set fees according to metered water use. In areas where sewer districts serve houses that are not on public water supplies, the wastewater-treatment facilities often maintain withdrawal information about self-supplied residences for billing purposes. This information can be used to estimate withdrawals for certain areas and to develop per-capita use coefficients along with population and housing data. Per-capita uses also are estimated by State pollution-control agencies when designing wastewater lagoons. These design values, usually about 80 to 100 gallons per capita per day (gpcd), may provide a good estimate of domestic per-capita use. Other coefficients, such as gallons per day per household or gallons per day per household market value, are available from the literature and can be used to determine domestic self-supplied withdrawals in areas for which ancillary data on housing are available.

Domestic self-supplied withdrawals typically are from wells. The source for 99 percent of the domestic withdrawals in 1995 was ground water (Solley and others, 1998). Information on the use of surface water for domestic supply may be obtained from the State public health agency, the State agency responsible for permitting or water-use data collection, census housing data, or local knowledge of areas where surface water is used. Cisterns used to collect rainwater for domestic use may be considered a surface-water source. Springs are considered surface water in some States and ground water in others.

Industrial

Category Description

Industrial water use includes water used for such purposes as fabricating, processing, washing, diluting, cooling, or transporting a product. Water used for industrial purposes also may be incorporated into products or used for facility sanitation and maintenance. Industrial water users are businesses classified in the SIC codes under construction and manufacturing (see table 3) for the water-use compilation. Industrial water supplies may be derived from ground and surface sources (self-supplied industrial withdrawals) or provided by a public water supplier (industrial deliveries). Depending on water-quality requirements, supplies may be fresh or saline.

Data Elements

Mandatory

- Ground-water withdrawals, freshwater, by county
- Ground-water withdrawals, saline water, by county
- Surface-water withdrawals, freshwater, by county
- Surface-water withdrawals, saline water, by county
- Ground-water withdrawals, freshwater, by aquifer

Optional

- Deliveries from public supply, by county
- Consumptive use, freshwater, by county
- Consumptive use, saline water, by county
- Number of facilities, by county
- Reclaimed wastewater, by county
- Any data aggregated by hydrologic cataloging unit

Sources of Information

Information on industrial facilities and their water use may be obtained from the following list of sources. This list may not exhaust all possibilities for every State.

- State agencies that administer water rights, allocate water to users, or collect water-use data
- State agencies that issue permits for the discharge of water
- Health departments or public water suppliers—many industries receive treated water for sanitary uses

- Wastewater treatment facilities
- State department of labor
- State and county planning departments
- State directories of manufacturers
- Local chamber of commerce
- County assessors and zoning boards
- U.S. Census Bureau:
  <http://www.census.gov/>

  1997 Economic Census-Manufacturing-Industry Series (Geographic Area Series):

  county-level data; number of establishments/ employees by manufacturing industry; accessed through the American FactFinder, U.S. Census Bureau's online data source: *<http://factfinder.census.gov/home/saff/ main.html?_lang=en>*

  sector-specific reports by State: *http://www.census.gov/epcd/www/econ97.html*

  list of publications available for the manufacturing sector: *<http://www.census.gov/prod/www/abs/ 97ecmani.html>*

  list of U.S. Census Bureau contacts for manufacturing: *<http://www.census.gov/contacts/www/ c-manufa.html>*

- U.S. Environmental Protection Agency (USEPA): *<http://www.epa.gov/>*

  Water Discharge Permits, Permit Compliance System (PCS) database: *<http://www.epa.gov/enviro/html/pcs/ pcs_query_java.html>*

  Access to data regarding facilities holding National Pollutant Discharge Elimination System (NPDES) permits. Specify the facilities by using any combination of facility name, geographic location, SIC code, and chemicals.

- Dun and Bradstreet: *<http://www.dnb.com/us/index.asp>*

  Source for lists of manufacturing establishments that includes SIC codes and number of employees, among other information. This information is updated quarterly.

- Harris InfoSource: *<http://www.harrisinfo.com/servlet/ HIServlet?moduleName=session&methodName= getHomePage>*

  Source for purchasing lists of manufacturing establishments. This site includes limited free searching of their database and trial reports. Profile reports include primary and secondary SIC codes, number of employees, Web address of company, if available. The information is updated at least once a year.

Compilation Techniques

The types of industries and magnitude of withdrawals vary throughout the United States. Major industrial groups identified by SIC code include food and kindred products (SIC 2011–2099); paper and allied products (SIC 2611–2679); chemicals and allied products (SIC 2812–2899); petroleum refining and related industries (SIC 2911–2999); and primary metals industries (SIC 3312–3399). Although these types of industries historically have used the most water per facility, the most important industry in any given county may not fall into one of these five groups. It is important, therefore, to consider any industries outside these groups.

Withdrawals

There are three general approaches used to compile withdrawal data by industries:

1. Acquire site information and withdrawal data for individual industries, focusing on larger ones while striving for an adequate representation of the withdrawals in each county.

2. Acquire site information with ancillary data on employment or production, and estimate water withdrawals using water-use coefficients. These coefficients are usually in the form of usage in gallons per day per employee or per unit of product.

3. Combine the two approaches by acquiring site information and withdrawals for the larger industries and using these data to develop water-use coefficients for estimating withdrawals for the smaller industries. The accuracy of withdrawals derived by using water-use coefficients is always questionable, especially in the industrial category; however, coefficients developed using local sites may be more reflective of physical and economic conditions in an area than are nationally derived coefficients.

Withdrawal data may be acquired through telephone contacts or surveys sent out by cooperating agencies. The level of detail that is obtained depends on the amount of time available to collect industrial facility data and the ability of the people that are contacted to provide this information. Any relevant privacy issues that may affect both the acquisition and use of industrial data should be considered. Useful data to acquire include:

- Facility name, mailing address, physical plant facility address
- County
- Contact person's name, title, and telephone number

- Industry description or principal products
- SIC codes—primary and secondary
- Estimated annual quantity of product produced
- Total number of employees
- Number of ground-water sources, aquifer names, number and depth of well(s)
- Number of surface-water sources, names of streams or water bodies
- Latitude and longitude of wells and (or) intakes
- Maps of facility and water intakes
- Name of any public water-supply sources
- Amounts of water withdrawn (quantity per unit time) from each source
- Amount or percentage of total withdrawal that is saline ground or surface water
- Method of determining withdrawals (meters, other)
- Water-use breakdown for various purposes—cooling, processing, sanitary use, boiler feed, power generation, other
- Amount of water recycled or re-used
- Wastewater discharge—average amount or percentage
- Number of days operating each year
- Average number of hours operating each day
- Approximate age of the facility
- Need for maintaining confidentiality

When using lists provided by permitting agencies, it is wise to determine the minimum withdrawal rate for issuing permits or collecting surveys. If this minimum withdrawal rate is large, a significant number of small industries may not be surveyed.

Any data obtained from industrial facilities through surveys, questionnaires, or cooperator reports need to be checked for accuracy and correct units of measurement. Some of an industry's water may be purchased from a public water supply for sanitary use within the facility or if the quality of the water needs to be high, such as in the food and beverage industry. Estimates of the total water needs for facilities with a combination of sources need to be adjusted so that only the self-supplied withdrawals from surface- and ground-water sources are counted.

Water-Use Coefficients

In the absence of reported data, industrial withdrawals may be estimated using water-use coefficients and ancillary data such as numbers of employees, production volumes, or annual sales. The largest industries should be contacted to acquire actual water-use withdrawal data. These withdrawals then can be used to develop locally adjusted water-use coefficients to estimate withdrawals for smaller industries.

The use of coefficients to estimate industrial water use is imprecise because of the variability in factors affecting water use by industries. The specific processes, age of the facility, cost of the water and wastewater treatment, and amount of recycling all contribute to the amounts of water needed by an industry. These factors should be considered when using national coefficients or when developing and adjusting local coefficients for nonsurveyed industries.

Allocation of Withdrawals by Source

If water use is estimated on the basis of ancillary data, the results must be allocated by source—ground water and surface water, freshwater and saline water. The best approach is to acquire as much site-specific data as possible for the larger industries. Some facilities have both ground-water and surface-water sources and possibly a public water-supply source as well. Public water-supply deliveries should not be included in the industrial withdrawals. If site-specific data are not available, plotting the industries' locations on a map can help determine the probable sources. Many industries are located in specific areas because of ample supplies of water from rivers or productive aquifers. Maps showing the extent of surficial and bedrock aquifers can indicate whether industries are located over a major aquifer. Water-use permits (if required) also should indicate if the industry withdraws from a protected aquifer. Finally, comparison with other nearby users such as public water suppliers may help determine an industry's source of water.

For some States, the source of saline surface water is the ocean. Saline ground water is characteristic of some aquifers that may be described in ground-water reports. The USGS ground-water specialist in each State may be able to help identify areas noted for having saline ground water.

Irrigation

Category Description

Irrigation water use includes fresh or reclaimed water that is applied by an irrigation system to sustain growth in agricultural and horticultural vegetation (SIC 0111–0191 and 4971 in table 3). It also includes water that is applied for pre-irrigation, frost protection, chemical application, weed control, field preparation, crop cooling, harvesting, dust suppression, leaching of salts from the root zone, and conveyance losses. Activities such as irrigation of public and private golf courses (SIC 7992 and 7997), parks, nurseries, turf farms, cemeteries, and other landscape-watering uses also are included in the irrigation category. The irrigation category includes additional data on irrigated acres by type of irrigation system—sprinkler, surface (flood), and microirrigation.

Water that is purchased from or provided by a public-supply system for irrigation users, golf courses, parks, cemeteries, and landscaping irrigation is not included in the irrigation category. This water is included in public-supply withdrawals and may have been accounted for as a commercial delivery or public use.

For the 2000 compilation, States optionally may divide irrigation water use into crop irrigation and golf-course irrigation. In this case, only golf-course irrigation withdrawals are included in golf-course irrigation; all other landscape irrigation is included with crop withdrawals. If irrigation is divided in this manner, county-level totals for both crop and golf-course irrigation are mandatory where applicable, and total irrigation water withdrawals are calculated as the sum of the two subcategories. If irrigation withdrawals are not divided into crop and golf-course withdrawals, then all irrigation is reported as total withdrawals and may include both types.

Data Elements

Mandatory

- Ground-water withdrawals, freshwater, by county
- Surface-water withdrawals, freshwater, by county
- Total acres irrigated by sprinkler systems, by county
- Total acres irrigated by surface systems, by county
- Total acres irrigated by microirrigation systems, by county
- Total ground-water withdrawals, by aquifer

Optional

- Divided crop and golf-course withdrawals, by county
- Consumptive use, by county
- Conveyance loss, by county
- Reclaimed wastewater, by county
- Any data aggregated by hydrologic cataloging unit

Sources of Information

Information about irrigation withdrawals and acres irrigated for crops or golf courses may be obtained from the following list of sources. This list is not all inclusive, and USGS water-use project chiefs are encouraged to investigate any other potential sources of information available for their States.

- State agencies that administer water rights, allocate water to users, or collect water-use data
- U.S. Department of Agriculture (USDA): <http://www.usda.gov/>

1997 Natural Resource Inventory:
<http://www.nrcs.usda.gov/technical/NRI/1997/
national_results.html>

Farm Service Agency:
<http://www.fsa.usda.gov/pas/default.asp>

Cooperative State Research, Education, and Extension
Service (CSREES):
<http://www.csrees.usda.gov/>

State partners of the CSREES (includes agricultural
experiment stations and county cooperative extension
offices):
<http://www.csrees.usda.gov/qlinks/partners/
state_partners.html>

Local partners:
<http://www.csrees.usda.gov/Extension/index.html>

Natural Resource Conservation Service, National
Engineering Handbook, Part 652, Irrigation Guide:
<http://www.wcc.nrcs.usda.gov/nrcsirrig/
irrig-handbooks-part652.html>

- National Agricultural Statistics Service (NASS):
<http://www.usda.gov/nass/>

Links to NASS Web sites for each State:
<http://www.usda.gov/nass/sso-rpts.htm>

  State Web sites may include links to information
  such as agricultural statistics, publications (including annual bulletins), and general information on
  agriculture in that State; other State agricultural
  agencies; and agricultural census data.

Published Estimates Data Base (PEDB)—custom queries for agricultural statistics, down to the county level:
<http://www.nass.usda.gov:81/ipedb/>

  Data files for 1999 crop acreage, by county:
  <http://www.usda.gov/nass/graphics/county99/
  indexdata.htm>

- Census of Agriculture:
<http://www.nass.usda.gov/census/>

Links to portable document file (PDF) versions of 1997
Census of Agriculture Volume 1 tables (national, State,
and county data):
<http://www.nass.usda.gov/census/census97/volume1/
vol1pubs.htm>

Highlights from the 1997 census of agriculture, State or
county level:
<http://www.nass.usda.gov/census/census97/
highlights/ag-state.htm>

  text and graphics on primary agricultural products,
  including acres of crops harvested

Profiles of the 1997 census of agriculture, State or
county level:
<http://www.nass.usda.gov/census/census97/profiles/
ag-state.htm>

text and graphics on differences from previous years

1998 Farm & Ranch Irrigation Survey:
<http://www.nass.usda.gov/census/census97/fris/
fris.htm>

  State-level data on acres irrigated by irrigation
  method; application rates by irrigation method,
  source of water, and crop type

1998 Census of Horticultural Specialties:
<http://www.nass.usda.gov/census/census97/
horticulture/horticulture.htm>

  State-level data on acres for horticultural crops,
  including acres irrigated by source of water and
  acres irrigated by type of irrigation system

- Bureau of Reclamation, U.S. Department of the
Interior:
<http://www.usbr.gov/>

Annual operating plans for Federally owned dams and
reservoirs operated by irrigation or reclamation
districts:
<http://www.usbr.gov/gp/water/index.cfm>

AgriMet is a network of automated agricultural
weather stations located in irrigated agricultural areas
throughout the Pacific Northwest and Montana:

  Pacific Northwest Region:
  <http://mac1.pn.usbr.gov/agrimet/>

  Great Plains Region (Montana):
  <http://www.usbr.gov/gp/agrimet/index.cfm>

- U.S. Environmental Protection Agency (USEPA):
<http://www.epa.gov/>

Water Discharge Permits, Permit Compliance System
(PCS) database:
<http://www.epa.gov/enviro/html/pcs/
pcs_query_java.html>

Access to data regarding facilities holding National
Pollutant Discharge Elimination System (NPDES) permits. Specify the facilities by using any combination of
facility name, geographic location, SIC code, and
chemicals.

- Irrigation Journal—2000 annual irrigation survey:
v. 51, no. 1, January/February 2001, p. 12–41 (Irrigation Journal, 2001; no longer published)

- GolfServ.com:
<http://www.golfserv.com/apps/courses/search.asp>

Users can search the database for a listing of golf
courses in each State. Includes data on public and private courses, number of holes, location, and links to
each course's profile and Web site (if they have one).
Registration on the Web site (free) is required to access
specific course data.

- Golfcourse.com:
    <*http://www.golfcourse.com/search/custom.cfm*>
    Users can search the database for a listing of golf
    courses in each State. Includes information on number
    of holes. No registration is required.

- Federal or State crop and livestock reporting services

- State and local turf-grower associations

- County assessor

- Land-grant universities—college of agriculture,
    departments of watershed science, soil science, plant
    science, crop science, or irrigation engineering

- Water management districts, irrigation districts,
    irrigation companies

- Irrigation equipment dealers

- Farm and crop improvement associations

- Individual golf courses, parks, and other recreational
    areas

- Professional, technical, and trade journals

- State departments of commerce and tourism

Compilation Techniques

For the irrigation category, information on withdrawals
from ground and surface sources is compiled for the counties
where the withdrawals occurred. Information on acres irrigated
is compiled for the counties where the irrigation water was
applied. In most cases, irrigation water is applied close to
where it is withdrawn; in some cases the irrigation water is
transported long distances through canals before being applied
to crops.

Direct methods for compiling information on irrigation
withdrawals and irrigated acres include use of reported data,
surveys, and personal contact. Indirect approaches for estimating withdrawals include calculation of crop water needs and statistical sampling. Data on irrigated acres by crop type and irrigation system type are used in most estimation methods for
determining withdrawals. Total withdrawals need to be disaggregated to ground- and surface-water sources if information on
site-specific withdrawals is not available. Compilation of withdrawals for golf-course irrigation is optional.

Reported Data

Some States require individual water users or water-right
holders to measure and report their withdrawals and irrigated
acreages for each well or surface-water diversion. These data
are considered the best because the location of the withdrawals
can be accurately assigned and aggregated to a county, aquifer,
or other area. Other sources of reported data may yield partial
coverage of irrigation use. USGS cooperative studies may
include measurement of site-specific data in local areas within

a State. Irrigation districts or other agencies may own water
rights and distribute water to users; these entities usually measure both withdrawals and deliveries. Water masters may have
been assigned to measure or compile measured withdrawals and
deliveries in some areas, in which case they are excellent
sources of data.

Reported measurements are the most defensible data; however, the completeness of reported data varies among States.
Before using data that are reported to or by other agencies, it is
important to understand the level of completeness, the degree of
accuracy, and the amount of QA/QC given to data obtained
from other Federal, State, and local agencies. USGS water-use
project chiefs need to verify whether the data are based on full
and complete enumerations, like a census, or are statistically
based, like the NASS data. Reported crop acreages may include
all crop acreages or just irrigated acreages. Irrigated acreages
and water deliveries may be reported for an area of management
of a particular agency, or for an entire county. Some data may
be censored for privacy reasons in areas with few irrigators.
Reported withdrawals and acreage data need to be reviewed for
errors, either by the reporting agency or by the USGS water-use
project chief. Knowing the level of QA/QC for data that are
reported by other agencies may prevent calculations that are
based on incorrect assumptions.

Surveys and Personal Contact

Another method of estimating irrigation water use is by a
local survey. Surveys that are conducted by USGS rather than a
cooperator need to comply with the requirements of the 1995
Paperwork Reduction Act. Ideally, survey forms include a complete background about why the information is being requested,
how it will be used, and by whom. Other important components
of the survey are (1) a concise description of the requested data,
(2) a contact name and phone number to direct questions to, and
(3) a desired completion date. In any case, efficient collection
and processing of survey data can best be achieved if the survey
forms are short and easy for the user to complete. Typically,
other State and local agencies distribute surveys, and the data
are shared with USGS.

It is important to get accurate data on the largest users, who
typically have more data available than small water users. Data
sources such as the USDA's Farm and Ranch Irrigation Survey
can be used to target counties with the largest water users when
developing survey lists. Some of the most beneficial survey
information includes:

- Total irrigated acreage

- Crop and pasture acreage

- Type and efficiency of irrigation system(s)

- Quantity or flow rate of water by source

- Irrigation scheduling and frequency

- Number of irrigation wells

- Total depth of well(s)

- Capacity of well(s)
- Contributing aquifer system(s)
- Total annual energy usage
- Power consumption coefficient(s) if known

Other useful data include information on crop water shortages, acres harvested and yields by crop, energy sources, water and agricultural management practices, and any agricultural resources that generally are contacted when the irrigator needs additional guidance.

Estimates of Crop Water Needs

A commonly used method to estimate irrigation withdrawals involves calculating consumptive use for irrigated crops using crop water-consumption coefficients for various crops and system types (see Blaney and Criddle, 1950; U.S. Department of Agriculture, 1970, 1976, 1997). The amount of water consumed by crops plus additional water used in conveyance or needed for other irrigation uses is the total withdrawal. This method requires that ancillary data exist for total irrigated acres for each type of crop, irrigation system efficiencies, conveyance losses, climatic variables, and other irrigation management practices such as pre-irrigation, frost protection, weed control, and leaching salts from soils. This "consumptive-use" technique assumes that the irrigation water applied is adequate for optimal plant growth and that the plants are not being irrigated with more or less water than needed.

Application efficiency is a measure of the effectiveness of the irrigation system in applying the right amount of water to the soil and root zones over time. Application efficiency must be taken into account if irrigation withdrawals are estimated from crop water-consumption coefficients. Application efficiencies vary with the type of irrigation system and soil, crop, topographic, and climatic conditions. Pertinent climatic conditions that have large effects on irrigation efficiencies include wind speed, relative humidity, and air temperature.

In many cases, compilation of State withdrawals on the basis of crop consumptive use plus additional climatic and soil factors is not feasible due to the amount of work and level of detailed data necessary. Use of previously determined crop consumption values may be the most cost-effective method of estimating irrigation withdrawals. These values may be obtained from sources such as the USDA irrigation guide (U.S. Department of Agriculture, 1970).

Statistical Sampling

For large irrigated areas where few measurements exist, statistical sampling represents a cost-effective way to estimate irrigation withdrawals. Withdrawals are measured at a set of sample sites where data for predictor variables such as power consumption, lift, or crop type also are known. Water-use coefficients are developed from these sample data and then are used to estimate withdrawals at unsampled sites where the predictor variable(s) are known.

Statistical approaches may have transfer value to other areas if there is good understanding of the predictor variables, their statistical significance, and the level of expected accuracy. For instance, by understanding the variability of the predictor variable, a specific number of sample sites may be determined that will enable calculations of withdrawals for all sites with a probability that the calculated values are within a specific margin of error. A complete description of how to determine a sufficient sample size is provided in Luckey (1972) and Helsel and Hirsch (1995).

Allocation of Withdrawals by Source

If site-specific measurements of withdrawals are unavailable, then the estimates of total withdrawals need to be allocated between ground- and surface-water sources. Geographic location is the greatest determinant of water availability from each source. The availability, both hydrologically and politically, of various aquifers and streams within each State is key to assigning source of water to estimates of irrigation withdrawals. Often, there is a predominant source of supply for a given State or geographic area within a State. For example, Midwestern States that overlie the High Plains aquifer use mostly ground water for irrigation. Western States, such as California, Idaho, Oregon, and Washington, use predominantly surface water for irrigation.

The best sources of information on sources of water for irrigation are State agencies and county extension agents that deal with the political and managerial aspects of water resources. Irrigation districts and equipment dealers are other good sources of local information. Patterns of use for other categories may suggest a reasonable division.

Acres Irrigated by System Type

Estimates of irrigated acres by system type are compiled for each State at the county level. Irrigated acreage is reported by three general methods of application—sprinkler, surface, and microirrigation. Many types of irrigation systems are included in each of these categories.

- Sprinkler methods include all boom, center-pivot, lateral-move, low-energy precision application (LEPA), permanent, portable, side-move, side-roll, solid-set, traveling-gun, towed, and other sprinkler irrigation systems.
- Surface methods include all borders, ditch, flood, furrow, gated-pipe, surge-flow, water-spreading, and other gravity systems;
- Microirrigation methods include all bubbler, drip, micro-jet, mist, porous trickle-tubing, spray, trickle, and other low-volume irrigation systems, and subsurface systems.

Reliable data on irrigated acreage also are essential for most methods of estimating irrigation withdrawals. Site-specific irrigated acres by irrigation system type are difficult to obtain except where State agencies require irrigators to report detailed information on crop types, acreages, and irrigation methods. The USDA census of agriculture, conducted in years ending in "2" and "7", provides the most recent national data set of irrigated crop acreages by county (U.S. Department of Agriculture, 1999a). The USDA farm and ranch irrigation survey, last released in 1998, provides a national data set of irrigated acreage by irrigation method by State (U.S. Department of Agriculture, 2000c). The Irrigation Journal for 2000 (now discontinued) also provides State-level irrigated acreages by method (Irrigation Journal, 2001). County assessors also are potential sources for county-level information on irrigated acreage and irrigation system type. Remote sensing has been utilized as an indirect method of determining acres irrigated and crop types (Raymond and others, 1992).

In some parts of the United States, the growing season is long enough that double and triple cropping on the same irrigated acreage can occur. In these cases, irrigated acres are accumulated to reflect the total acreage irrigated during the year; therefore, when any acre of land is cropped twice, it is counted as 2 irrigated acres. Counting acreage in this manner produces a valid application rate, expressed in terms such as acre-foot per acre. The irrigation method used on the subsequent crops also may vary and should be investigated and documented for each State. If multiple irrigation methods are used on a crop in a single growing season, acreages are reported under the method that provides the majority of water to the crop.

Golf-Course Irrigation

Golf courses may utilize water from ground or surface sources, purchased water (from a public supplier or irrigation district), reclaimed water from a public wastewater treatment facility, or a combination of these sources. For golf courses, the best information on source of water is obtained directly from the golf-course maintenance personnel. These people usually know how much water is diverted, withdrawn, or delivered from each of the possible sources.

Factors affecting the amount of irrigation water used at golf courses include course design, climatic conditions, acreage, irrigation systems, soils, availability of water for irrigation, and local irrigation practices. Normally, more water per unit of area is applied to the greens and tees than to the fairways. It is best to obtain metered withdrawals for specific golf courses. If site-specific data are not available, irrigation withdrawals may be estimated using coefficients developed from surveys of golf courses in a State. A survey should include all of the information necessary to compute an application rate on the basis of the consumptive needs of the grasses and the irrigated acres.

Thermoelectric Power

Category Description

Thermoelectric power water use is defined as the amount of water used in the process of generating thermoelectric power. All thermoelectric-power generation facilities are classified as SIC 4911 (table 3). The source of the power may be fossil fuels, nuclear fission, or geothermal energy. Thermoelectric power-plants typically generate electricity with a boiler, where water is heated to turn it into steam. The steam then is used to turn turbines, which generate electricity. After the steam is used to turn the turbines, the steam is condensed to water by cooling it in a heat exchanger. The condensed water then is routed back to the boiler, where the cycle begins again. The predominant use of water is to cool the steam.

Water withdrawal requirements at power-generation facilities depend primarily on whether or not the cooling water is recirculated. The two general types of cooling are once-through (open-loop) cooling and closed-loop cooling. For the 2000 compilation, water withdrawal estimates are compiled for each type of cooling system. Once-through or open-looped cooling requires the largest amounts of water withdrawal because it is not recirculated within the facility. The water is withdrawn from a source, circulated through the heat exchangers, and then returned to a water body at a higher temperature. This technology is common in older facilities but generally is not used for new facilities because of increasingly restrictive thermal requirements for return water.

Close-looped cooling systems utilize cooling ponds and cooling towers to recirculate water within the system, thus reducing the overall water withdrawal requirement. Withdrawals to replace cooling water lost to evaporation, blowdown, drift, and leakage are considered "makeup" water. A cooling pond is a shallow reservoir with a large surface area to remove heat from circulation water. The rate of heat loss may be enhanced through the use of spray nozzles. A cooling tower is a structure designed to remove heat from water. The heated circulation water is sprayed into the tower and is cooled by radiation from the sides of the tower or contact with the cooler air. Cooling towers commonly are used where land and water are expensive, or where local regulations prohibit the release of thermal water.

Some industrial facilities also generate thermoelectric power. These facilities are called "cogeneration" facilities. If the data provided from the industrial facility are sufficient to identify water used in power generation, then that amount of water is reported separately as a power generation water use.

Data Elements

Mandatory

- Ground-water withdrawals, once through, freshwater, by county
- Surface-water withdrawals, once through, freshwater, by county
- Surface-water withdrawals, once through, saline water, by county
- Ground-water withdrawals, closed loop, freshwater, by county
- Surface-water withdrawals, closed loop, freshwater, by county
- Surface-water withdrawals, closed loop, saline water, by county

Optional

- Deliveries from public supply, once through, by county
- Deliveries from public supply, closed loop, by county
- Consumptive use, once through, freshwater, by county
- Consumptive use, once through, saline water, by county
- Consumptive use, closed loop, freshwater, by county
- Consumptive use, closed loop, saline water, by county
- Power generation, once through, by county
- Power generation, closed loop, by county
- Number of facilities, once through, by county
- Number of facilities, closed loop, by county
- Reclaimed wastewater, once through, by county
- Reclaimed wastewater, closed loop, by county
- Any data aggregated by hydrologic cataloging unit

Sources of Information

Information concerning thermoelectric power withdrawals may be obtained from the following list of sources. This list is not all inclusive, and USGS water-use project chiefs are encouraged to investigate any other potential sources of information available for their States.

- State agencies that administer water rights, allocate water to users, or collect water-use data
- Individual facilities
- Public suppliers (for deliveries)
- U.S. Department of Energy (DOE):
  <http://www.doe.gov>
  Energy Information Administration (EIA, part of DOE):

  <http://www.eia.doe.gov/>
  List of EIA electric power forms:
  <http://www.eia.doe.gov/oss/forms.html>
  This list includes contact information for each type of report.

  EIA–412 (Annual Report of Public Electric Utilities)—generating plant data by type of plants; for municipal and Federally owned electric utilities:
  <http://www.eia.doe.gov/cneaf/electricity/page/eia412.html>

  EIA–906 (Monthly Power Plant Report, formerly EIA–759)—net generation, fuel type; monthly data for plants with nameplate capacity of 50 megawatts or more, annual data for plants with nameplate capacity of less than 50 megawatts:
  <http://www.eia.doe.gov/cneaf/electricity/page/eia906u.html>

  EIA–767 (Annual Steam-Electric Plant Operation and Design Data)—net generation, fuel type, cooling water process/source/rate; only plants with generating capacity of 10 or more megawatts:
  <http://www.eia.doe.gov/cneaf/electricity/page/eia767.html>

  Data matrix of EIA electric power forms online:
  <http://www.eia.doe.gov/cneaf/electricity/forms/datamatrix.html>

  Energy data for each State (summaries, profiles, graphics):
  <http://www.eia.doe.gov/emeu/states/_states.html>

    Electricity profile contains a graphic with largest investor-owned utilities in the State and largest plants in the State; 1998 data.

    Electricity restructuring plans contain links to State regulatory commissions and major utilities.

- U.S. Environmental Protection Agency (USEPA):
  <http://www.epa.gov/>

  Water Discharge Permits, Permit Compliance System (PCS) database:
  <http://www.epa.gov/enviro/html/pcs/pcs_query_java.html>

    Access to data regarding facilities holding National Pollutant Discharge Elimination System (NPDES) permits. Specify the facilities by using any combination of facility name, geographic location, SIC code, and chemicals.

- State agency responsible for compliance with USEPA's Clean Water Act Program
- State agency for power administration
- Regional "power pools" (groups of electric utility companies)

Compilation Techniques

Most water for thermoelectric use is self supplied from fresh or saline surface-water sources. Smaller quantities are derived from ground-water sources or provided by public suppliers.

Information collected and maintained by the U.S. Department of Energy, Energy Information Administration (EIA), includes powerplant ownership, location, cooling method, sources of water, average withdrawal rates, average discharge rates, operating status, and power generated. This information is collected from monthly and annual surveys of powerplants. Information on power generation and energy source may be found in the database EIA–906 Monthly Power Plant Report, formerly EIA–759. EIA–767, Annual Steam-Electric Plant Operation and Design Data, contains cooling information for all organic- and nuclear-powered plants with a generator name-plate rating of 10 or more megawatts.

Information on the withdrawal of water for the process of generating electric power also can be obtained from each utility. If the contact person for a utility is not known, a good place to start is the person at the utility who prepares the Discharge Monitoring Reports (DMR's) for USEPA. DMRs contain information on volume discharged from all pipes in the facility and can be compared to the permit or permit application as to the source of the water and how it was used. The following selected terminology may help in conversations with utility personnel—blowdown, capacity, drift, and makeup water. These terms are defined in the "Glossary."

State agencies that administer water rights or monitor water use may provide more detailed information on actual withdrawals for thermoelectric power generation. The State agency that is responsible for compliance with USEPA's Clean Water Act is an important source for this information. USEPA administers the Permit Compliance System (PCS) database, which was designed to track permit, compliance, and enforcement status data for the National Pollutant Discharge Elimination System (NPDES) Program under the Clean Water Act. An NPDES permit is required for all point discharges into United States waterways. The PCS database contains descriptive information on major power-generating facilities, their location, and monthly return flows. The NPDES permit application and the permit itself usually include detailed descriptions of the plant that provide basic information on all the sources of supply for the plant, the different ways in which water is used in the plant, and water included in the reported discharge values.

Power generation data can be used to estimate thermoelectric water use if it is not possible to obtain withdrawal amounts from other sources. A coefficient to estimate the gallons of water used per kilowatt-hour of electricity generated is calculated using information on withdrawals and power generation from plants of similar age, heating processes, and cooling methods. This coefficient then can be multiplied by the amount of electricity generated over a specified time period by the plant for which withdrawals are being estimated. Power generation data are available on EIA–906 on form 759; however, some of the electricity generated at a thermoelectric powerplant is used to run the powerplant itself. The net amount of power produced generally is reported. Water withdrawals ideally are calculated using the gross power produced, if available.

Mining

Category Description

Mining water use includes water that is used for the extraction of minerals, ores, and gases, which may be in the form of a solid, liquid, or gas. Solid minerals include coal, ores, sand, and gravel. Liquids pertain to crude petroleum, and gases mostly pertain to natural gas. The mining water-use category includes quarrying, milling (crushing, screening, washing, and flotation), and other operations as part of a mining activity. Water that is pumped and then re-injected for secondary oil recovery is considered a water use and should be included. Water pumped from mines to dewater them, or water that is produced as a by-product of primary oil production is not included in this category. That is, the water is ignored if it is drained and discharged without being put to use, even though the water is transferred from ground water to surface water. If water is put to a beneficial use such as washing or dampening roads for dust control, then the water is included in mining water-use estimates.

Four major SIC groups account for most of the water used in the mining category (see table 3). These major groups are metal mining (SIC 1011–1099), coal mining (SIC 1221–1241), oil and gas extraction (SIC 1311–1389), and mining and quarrying of nonmetallic minerals, except fuels (SIC 1411–1499). The mining water-use category does not include the processing of raw materials, such as smelting ores, refining petroleum, and slurry pipeline operations. These are considered industrial uses of water.

Data Elements

Compilation of estimates for the mining category is mandatory for Alaska, Arizona, California, Florida, Indiana, Minnesota, Nebraska, Oklahoma, Pennsylvania, Texas, Utah, and Wyoming. Other States are encouraged to compile data on mining water use if mining activity is important to their State.

Mandatory

- Ground-water withdrawals, freshwater, by county
- Ground-water withdrawals, saline water, by county
- Surface-water withdrawals, freshwater, by county
- Surface-water withdrawals, saline water, by county

Optional

- Consumptive use, freshwater, by county

- Consumptive use, saline water, by county
- Reclaimed wastewater, by county
- Any data aggregated by hydrologic cataloging unit

Sources of Information

Information concerning mining water use may be obtained from the following list of sources. This list is not all inclusive, and USGS water-use project chiefs are encouraged to investigate any other potential sources of information available for their States.

- Individual facilities (by personal visits and/or cooperator surveys)
- State agencies that issue permits for the use of water
- State agencies that issue permits for land reclamation or erosion control
- State agencies that issue permits for mining activities
- State agencies for economic development
- Universities, departments of geology and mines
- U.S. Census Bureau:
  <http://www.census.gov/>
  1997 Economic Census–Mining–Industry Series (Geographic Area Series)

    contains State-level data only (mining industry data not published by county); number of establishments/employees by mining industry:
    <http://www.census.gov/epcd/www/97EC21.HTM>

    contains a list of publications available for the mining sector:
    <http://www.census.gov/prod/www/abs/97ecmini.html>

    contains a list of U.S. Census Bureau contacts for mining:
    <http://www.census.gov/contacts/www/c-manufa.html>

- Office of Surface Mining, U.S. Department of the Interior:
  <http://www.osmre.gov/links.htm>
  Links to a variety of Web resources including selected State departments responsible for regulation and abandoned mine land programs, Federal agencies, interest groups, and selected State mining and reclamation associations.
- U.S. Geological Survey, Department of the Interior:
  <http://www.usgs.gov>
  State minerals statistics and information:
  <http://minerals.usgs.gov/minerals/pubs/state/>
  Links to minerals information for each State, including the Minerals Yearbook. Also available for each State are maps showing major mineral-producing areas by

county, contacts, and links to State-specific minerals agencies.
Mine and mineral processing plant locations:
<http://minerals.usgs.gov/minerals/pubs/mapdata/>
Accessible from this page are maps with plant/mine locations, GIS-ready data layers, and tables of plant/mine county locations by State and commodity.

- Bureau of Land Management, U.S. Department of Interior:
  <http://www.blm.gov/nhp/index.htm>
  Manages public lands, primarily in the western United States. Each State office has a mineral section.

    The Land and Mineral Records—LR2000 system:
    <http://www.blm.gov/lr2000/>

- National Mining Association:
  <http://www.nma.org/default.asp>
  Coal Producer Survey:
  <http://www.nma.org/statistics/pub_coal_survey.asp>
  Tables of major underground and surface coal mines in the United States; includes name of mine, State, and operating company.
- Association of American State Geologists:
  <http://www.kgs.ukans.edu/AASG/AASG.html>
  Links to each State's Geological Survey.
- Mining technology, Web site for the mining industry:
  <http://www.mining-technology.com/industry/united_states.html>
  Links for commodity and State agencies.
- U.S. Environmental Protection Agency (USEPA):
  <http://www.epa.gov/>

    Water Discharge Permits, Permit Compliance System (PCS) database:
    <http://www.epa.gov/enviro/html/pcs/pcs_query_java.html>

    Access to data regarding facilities holding National Pollutant Discharge Elimination System (NPDES) permits. Specify the facilities by using any combination of facility name, geographic location, SIC code, and chemicals.

Compilation Techniques

Withdrawals by mining facilities vary depending on the size and type of mining operation. Collection of site-specific data is the best approach. If these data are not reported to a State agency or collected by survey, then the largest mines in each State should be contacted individually to determine the quantity and source of water used for mining purposes. State or local agencies also may have useful information in water permit files, although often permitted amounts are different from actual water use.

In some States, surveys are used to compile information for State Geological Survey annual mineral reports. These data may provide information on total production for State mining operations, by mineral or commodity; however, they are not always reported at the county level. Nonetheless, if good water-use coefficients are available, statewide total water-use estimates may be derived using State production data and county totals disaggregated using local information.

It may be possible to develop water-use coefficients for certain types of mining using site-specific data from facilities with similar operations. Coefficients usually are based on total water use and production values. For example, they may be expressed in terms of the amount of water used to produce a specific amount of ore (millions of gallons per ton), or water used to generate a volume of sales (millions of gallons per unit of annual revenue). The U.S. Census Bureau economic census (conducted every 5 years, in years ending with 2 and 7) provides State-level information for various types of mining establishments, including number of establishments, number of employees, value of shipments, production, and payroll (U.S. Census Bureau, 1999). Coefficients should not be applied without investigating the local water-use and mining practices in each State.

Livestock

Category Description

Livestock water use is water associated with livestock watering, feedlots, dairy operations, and other on-farm needs. Livestock includes dairy cows and heifers, beef cattle and calves, sheep and lambs, goats, hogs and pigs, horses, and poultry. Poultry includes chickens, turkeys, ducks, geese, pheasants, and pigeons. Livestock water use also includes dairy and poultry sanitation or wash down, waste-disposal systems, cooling of animals and products, and incidental water losses. Livestock and livestock products are classified as SIC 0211–0272, 0279, and 0291 (table 3). Animal specialties such as horses and other equines and fur-bearing animals are included in the livestock category for the 2000 compilation. In the 1990 and 1995 compilations, animal specialties water use was a separate subset of the livestock category, and also included fish-farm water use (Solley and others, 1993, 1998). Fish-farm and fish-hatchery water use are included in the aquaculture category for 2000. The livestock water-use category does not include on-farm domestic use, lawn and garden watering, or irrigation water use.

Data Elements

Compilation of estimates for the livestock category is mandatory for California, Iowa, Kansas, Minnesota, Missouri, Nebraska, North Carolina, Oklahoma, Texas, and Wisconsin.

Other States are encouraged to compile data on livestock water use if possible.

Mandatory

- Ground-water withdrawals, freshwater, by county
- Surface-water withdrawals, freshwater, by county

Optional

- Consumptive use, freshwater, by county
- Any data aggregated by hydrologic cataloging unit

Sources of Information

Information concerning livestock water use and the number of animals and farms may be obtained from the following list of sources. This list is not all inclusive, and USGS water-use project chiefs are encouraged to investigate any other potential sources of information available for their States.

- State agencies that administer water rights, allocate water to users, or collect water-use data

- U.S. Department of Agriculture (USDA):
  <http://www.usda.gov/>

  Cooperative State Research, Education, and Extension Service (CSREES):
  <http://www.csrees.usda.gov/>

  State partners of the CSREES (includes agricultural experiment stations and county cooperative extension offices):
  <http://www.csrees.usda.gov/qlinks/partners/state_partners.html>

  Local partners:
  <http://www.csrees.usda.gov/Extension/index.html>

  Economics, Statistics, and Market Information System:
  <http://jan.mannlib.cornell.edu/>

    Includes reports and data sets on livestock, dairy, and poultry

- U.S. Environmental Protection Agency (USEPA):
  <http://www.epa.gov/>

  Water Discharge Permits, Permit Compliance System (PCS) database:
  <http://www.epa.gov/enviro/html/pcs/pcs_query_java.html>

    Access to data regarding facilities holding National Pollutant Discharge Elimination System (NPDES) permits. Specify the facilities by using any combination of facility name, geographic location, SIC code, and chemicals.

- National Agricultural Statistics Service (NASS):
  <http://www.usda.gov/nass/>

Links to NASS Web sites for each State:
<*http://www.usda.gov/nass/sso-rpts.htm*>

A State's Web site may include links to information such as agricultural statistics, publications (including annual bulletins), and general information on agriculture in that State, other State agricultural agencies, and census data for that State.

Published Estimates Data Base (PEDB)—allows custom queries for agricultural statistics, down to the county level:
<*http://www.nass.usda.gov:81/ipedb/*>

1997 census of agriculture:
<*http://www.nass.usda.gov/census/*>

Links to PDF versions of 1997 census of agriculture, volume 1 tables (national, State, and county data):
<*http://www.nass.usda.gov/census/census97/volume1/vol1pubs.htm*>

Highlights from the 1997 census of agriculture, State or county level:
<*http://www.nass.usda.gov/census/census97/highlights/ag-state.htm*>

text and graphics on primary agricultural products

Profiles of the 1997 census of agriculture, State or county level:
<*http://www.nass.usda.gov/census/census97/profiles/ag-state.htm*>

text and graphics on differences from previous years

- Agricultural Network Information Center:
<*http://laurel.nal.usda.gov:8080/agnic/*>

- AgriSurf —The Farmer's Surf Engine:
<*http://www.agrisurf.com/*>

Search for livestock-related links, by topic (such as feedlots or specific types of livestock), especially for information on a specific business:
<*http://www.agrisurf.com/agrisurfscripts/agrisurf.asp?R=wnghudp&index=_25_26&c=World&d=World&DT=0&FG=0&DL=0&ur=0*>

- Federal or State crop and livestock reporting services

- Agricultural universities or colleges—departments of animal science

- Professional, technical, and trade journals

- State and county cooperative extension service agents

Compilation Techniques

Livestock water withdrawals generally are estimated by multiplying the animal count for each kind of livestock in a State by the respective per-capita use coefficient for each type of animal. Annual statistics on numbers of animals are available from the USDA National Agricultural Statistics Service as well as various State and county offices. Authorized or reported water use by individual facility may be available in some States that issue permits for large users such as feedlots or confined animal operations.

Livestock water use includes water used for drinking, sanitation, and washing. Water-use coefficients for animals vary with climate and local animal-husbandry practices and are based on the conditions in each State. Per-head use coefficients for various types of livestock are obtained from State water-permitting agencies, county extension service agents, or university agriculture departments. For some types of livestock, different coefficients are used depending on the age of the animal. For example, coefficients for cattle are larger than for calves, and coefficients for hogs may be larger than for pigs. Coefficients also depend on the purpose of the livestock. For example, dairy cows require more water per head than beef cattle because of increased stall-washing requirements for dairy operations. Another determining factor in per-head water-use coefficients is the nature of the facility. Indoor confined animal-feeding operations require more water for washing and waste disposal than outdoor operations. It may be necessary to collect additional information on volumes of water used at large facilities for comparison with commonly used coefficients for a State.

Information on source of water for livestock may be obtained from State agencies responsible for issuing permits for water use, county extension service agents, or personnel at university agriculture departments. Local knowledge of water availability from farm ponds or wells also can be used to develop percentages of ground and surface water used for livestock in each county. Allocation of estimated livestock withdrawals to ground- and surface-water sources can be difficult, particularly if each source is used at a different time of year or if climatic conditions affect the availability of either source throughout the year.

Aquaculture

Category Description

Aquaculture water use includes farming of animals that live in water, such as finfish and shellfish, for food, restoration, conservation, or sport. Aquaculture production occurs under controlled feeding, sanitation, and harvesting procedures primarily in ponds, flow-through raceways, and, to a lesser extent, in cages, net pens, and closed-recirculation tanks. This category includes both fish farms (SIC 0273) and fish hatcheries (SIC 0921), as shown in table 3. For the 1990 and 1995 compilations, fish-farm water use was included in the animal specialties subset of the livestock category, and fish-hatchery water use was included in the commercial category (Solley and others, 1993, 1998). For the 2000 compilation, water use by both fish farms and fish hatcheries is included in the aquaculture category.

Data Elements

Compilation of estimates for the aquaculture category is mandatory for Alabama, Arkansas, California, Idaho, Louisiana, Mississippi, North Carolina, and Utah. Other States are encouraged to compile data on aquaculture water use if possible.

Mandatory

- Ground-water withdrawals, freshwater, by county
- Surface-water withdrawals, freshwater, by county

Optional

- Ground-water withdrawals, saline water, by county
- Surface-water withdrawals, saline water, by county
- Consumptive use, freshwater, by county
- Consumptive use, saline water, by county
- Any data aggregated by hydrologic cataloging unit

Sources of Information

Information on aquaculture water uses may be obtained from the following list of sources. This list is not all inclusive, and USGS water-use project chiefs are encouraged to investigate any other potential sources of information available for their States.

- State agencies that administer water rights, allocate water to users, or collect water-use data
- U.S. Department of Agriculture, National Agricultural Statistics Service (NASS): <http://www.usda.gov/nass/>

  1998 census of aquaculture: <http://www.nass.usda.gov/census/>

  State-level data, in PDF format tables: <http://www.nass.usda.gov/census/census97/aquaculture/aquaculture.htm>

- U.S. Department of Agriculture: <http://www.usda.gov/>

  Animal and Plant Health Inspection Service (APHIS), National Aquaculture Program, Aquaculture Industry Reports: <http://www.aphis.usda.gov/vs/aqua/aquaindu.html>

  Economic Research Service, Aquaculture Outlook: <http://usda.mannlib.cornell.edu/reports/erssor/livestock/ldp-aqs/>

- U.S. Fish and Wildlife Service: <http://www.fws.gov/>
  List of U.S. Fish and Wildlife offices in each State or territory: <http://offices.fws.gov/directory/listofficemap.html>

National Fish Hatchery System: <http://fisheries.fws.gov/FWSFH/NFHSmain.htm>
Lists of hatcheries by State and region: <http://fisheries.fws.gov/FWSFH/NFHmapg.htm>

U.S. Environmental Protection Agency (USEPA): <http://www.epa.gov/>

National Agriculture Compliance Assistance Center: <http://www.epa.gov/agriculture/>

  Aquaculture operations: <http://www.epa.gov/agriculture/anaquidx.html>

  Production and general information: <http://www.epa.gov/agriculture/anaqupro.html>

Water Discharge Permits, Permit Compliance System (PCS) database: <http://www.epa.gov/enviro/html/pcs/pcs_query_java.html>

  Access data regarding facilities holding National Pollutant Discharge Elimination System (NPDES) permits. Specify the facilities by using any combination of facility name, geographic location, SIC code, and chemicals.

- AgriSurf—The Farmer's Surf Engine: <http://www.agrisurf.com/agrisurfscripts/agrisurf.asp?index=_25>

  Search for aquaculture-related links, especially for information on a specific business: <http://www.agrisurf.com/agrisurfscripts/agrisurf.asp>

- Aquaculture Network Information Center (AquaNIC): <http://aquanic.org/>

- Aquatic Network: <http://www.aquanet.com/>
  List of aquaculture associations: <http://www.aquanet.com/resources/aqua/aq_assn2.htm>

- Agricultural Network Information Center: <http://laurel.nal.usda.gov:8080/agnic/>

- State departments of agriculture or natural resources, divisions of fish and game

- Agricultural universities or colleges—departments of fisheries and wildlife

- Private aquaculture companies

Compilation Techniques

Withdrawals for offstream fish hatcheries and fish farms may be difficult to estimate because the water is used only as a vehicle to raise or hold the fish, and large quantities of water are still available for other uses. Water use by fish hatcheries and fish farms often is equivalent to the amount of water necessary to maintain pond water levels to offset losses due to leakage and evaporation. In other cases, the water is withdrawn from a stream channel and flows through the facility before returning to the stream channel. Evaporative losses are minimal in these flow-through systems. It is best to obtain local information on sources of water and volumes used by fish-hatchery and fish-farm operations. Coefficients to estimate water use by fish may be expressed in terms of water volume per facility, water volume per pond surface area, or water volume or pond surface area per pound of organism produced.

Aquifer Withdrawals

Aggregation of data on ground-water withdrawals of freshwater by aquifer for the public-supply, industry, and irrigation categories is mandatory for the 2000 national water-use compilation. Ground-water withdrawals of saline water are not reported by aquifer.

An aquifer is a geologic formation, group of formations, or part of a formation that contains sufficient saturated permeable material to yield significant quantities of water to wells and springs. Data are aggregated for each State by the principal aquifers and aquifer systems identified in table 2. Aquifers and aquifer systems listed in table 2 are identified in the National Atlas of the United States and illustrated on the USGS principal aquifers map (U.S. Geological Survey, 2003). The aquifer list is not all inclusive; aquifers are chosen to represent those with great areal extent, substantial productivity or use, and (or) national significance. Many smaller aquifers have been included within a common classification, such as "alluvial aquifers." Withdrawals from aquifers known by local or regional names that are not included in table 2 are allocated to the appropriate principal aquifer or aquifer system using judgment and expertise in each State, or are assigned to the "other" aquifer category.

Sources of Information

Information needed to assign ground-water withdrawals to principal aquifers and aquifer systems may be obtained from the following list of sources.
* USGS Ground-Water Atlas:
  <http://capp.water.usgs.gov/gwa/gwa.html>
* USGS ground-water specialists or ground-water scientists
* USGS Regional Aquifer Systems Analysis (RASA) reports and maps (bibliography in Sun and others, 1997):
  <http://water.usgs.gov/ogw/rasa/html/introduction.html>
* USGS National Water-Quality Assessment (NAWQA) studies (links to summary reports, Web sites, and bibliography:
  <http://water.usgs.gov/nawqa/pubsmain.html>)

* Well information from the USGS National Water Information System Ground-Water Site Inventory (GWSI) database:
  <http://waterdata.usgs.gov/nwis/gw>
* University geology departments or hydrology programs
* State agencies, such as those that permit the use of water resources
* Professional, technical, and trade journals

Compilation Techniques

Ground-water withdrawals of freshwater by principal aquifer and aquifer system for each water-use category are best compiled with as much site-specific information as possible. The public-supply and industrial categories generally have the most site-specific data and provide the best opportunity to allocate withdrawals to specific aquifers or aquifer systems. If irrigation withdrawals are not derived from site-specific data, information may be obtained from the USGS Ground-Water Atlas, RASA maps and reports, or other local ground-water publications that show the areal extent of aquifers and aquifer systems, their depths and saturated thicknesses, productivity, and water quality. The USGS Ground-Water Atlas shows volumes of ground-water withdrawals of freshwater that were reported for 1995 and other years.

The preferred method of allocating public-supply withdrawals to principal aquifers is to use information from each utility or State agency that maintains well data. Site-specific data should be collected from utilities that supply the largest volume of water or that serve the largest number of people. Some public suppliers withdraw from multiple aquifers. Aquifers for utilities that serve smaller populations or geographic areas may be identified using maps showing the full areal extent of the most productive aquifers with adequate water quality. Consultation with State ground-water experts may be needed when assigning withdrawals by small public-supply facilities to aquifers using the map of full areal extent.

If site-specific data are used to compile industrial withdrawals, aquifer information should be collected at the same time as withdrawal information. If industrial withdrawals are estimated using other methods that are not based on site-specific data, then aquifers may be identified using maps showing the full areal extent of productive aquifers with adequate water quality. If the industry locations are known only at the county level or if there are multiple aquifers or aquifer systems that might provide water, then the best available information on principal aquifers used may be obtained from local ground-water reports and USGS ground-water specialists.

If site-specific data for irrigation withdrawals are available, then aquifers may be identified using well drillers' logs or with data provided by the State agency that issues water permits. If irrigation estimates are derived using other techniques, such as estimates of crop water needs, then ground-water withdrawals may be allocated using comparisons between irrigated areas and the full areal extent, depth to water, and productivity of principal aquifers and aquifer systems. RASA and other USGS ground-water reports, along with expertise and knowledge from USGS ground-water specialists and local irrigators, can be used to help allocate withdrawals to the appropriate aquifers.

If ground-water withdrawals are identified from an aquifer with a local or regional name that is not in table 2, the aquifer usually can be correlated to one of the principal aquifers or aquifer systems. Some of the aquifer names represent composite aquifer systems and include many individual smaller aquifers. Withdrawals from an aquifer that cannot be correlated properly to one of the principal aquifers or aquifer systems are reported using the "other" aquifer category.

Ground-water withdrawals of freshwater by aquifer for each State are reported by point of withdrawal rather than by point of use, although some water subsequently may be transferred for use in a different State from which it was withdrawn. Withdrawals should be reported for the State where the withdrawals occur even if the same aquifer underlies several States.

Aquifer withdrawals are to be reported at the State level. Total State aquifer withdrawals of freshwater for each category should not exceed the total ground-water withdrawals for that category. USGS water-use project chiefs are required to document the sources of information and methods used to allocate withdrawals to each aquifer or aquifer system.

Selected References

Arvin, D.V., 1992, Feasibility of using portable, noninvasive pipe flowmeters and time totalizers for determining water use: U.S. Geological Survey Water-Resources Investigations Report 91–4110, 65 p.

Blaney, H.F., Sr., and Criddle, W.D., 1950, Determining water requirements in irrigated areas from climatological and irrigation data: U.S. Department of Agriculture, Soil Conservation Service, SCS–TP–96, 48 p.

Bureau of Land Management, 2004, Land and mineral records: Information available on World Wide Web, accessed April 14, 2004, at http://www.blm.gov/lr2000/

Dash, R.G., 1999, Comparison of two approaches for determining ground-water discharge and pumpage in the lower Arkansas River Basin, Colorado, 1997–98: U.S. Geological Survey Water-Resources Investigations Report 99–4221, 39 p.

Helsel, D.R., and Hirsch, R.M., 1995, Statistical methods in water resources: New York, Elsevier, Studies in Environmental Science 49, 529 p.

Hurr, T.R., and Litke, D.W., 1989, Estimating pumping time and ground-water withdrawals using energy-consumption data: U.S. Geological Survey Water-Resources Investigations Report 89–4107, 27 p.

Hutson, S.S., Barber, N.L., Kenny, J.F., Linsey, K.S., Lumia, D.S., and Maupin, M.A., 2004, Estimated use of water in the United States, 2000: U.S. Geological Survey Circular 1268, 46 p.

Irrigation Journal, 2001, 2000 Annual irrigation survey: Irrigation Journal, v. 51, no. 1, January–February 2001, p. 12–41.

Kjelstrom, L.C., 1991, Methods of measuring pumpage through closed-conduit irrigation systems: Journal of Irrigation and Drainage Engineering, v. 117, no. 5, p. 748–757.

Luckey, R.R., 1972, Analyses of selected statistical methods for estimating ground-water withdrawal: Water Resources Research, v. 8, no. 1, p. 205–210.

MacKichan, K.A., 1951, Estimated use of water in the United States, 1950: U.S. Geological Survey Circular 115, 13 p.

MacKichan, K.A., 1957, Estimated use of water in the United States, 1955: U.S. Geological Survey Circular 398, 18 p.

MacKichan, K.A., and Kammerer, J.C., 1961, Estimated use of water in the United States, 1960: U.S. Geological Survey Circular 456, 26 p.

Maupin, M.A., 1999, Methods to determine pumped irrigation-water withdrawals from the Snake River between Upper Salmon Falls and Swan Falls Dams, Idaho, using electrical power data, 1990–95: U.S. Geological Survey Water-Resources Investigations Report 99–4175, 14 p. with appendices.

Murray, C.R., 1968, Estimated use of water in the United States, 1965: U.S. Geological Survey Circular 556, 53 p.

Murray, C.R., and Reeves, E.B., 1972, Estimated use of water in the United States, 1970: U.S. Geological Survey Circular 676, 37 p.

Murray, C.R., and Reeves, E.B., 1977, Estimated use of water in the United States, 1975: U.S. Geological Survey Circular 765, 37 p.

National Mining Association, 2004, Annual coal producer survey: Information available on World Wide Web, accessed April 14, 2004, at http://www.nma.org/statistics/pub_coal_survey.asp

Office of Management and Budget, 1987, Standard industrial classification manual, 1987: Washington, D.C., U.S. Government Printing Office, 705 p.

Raymond, L.H., Nalley, G.M., and Rettman, P.L., 1992, Evaluation of the use of remote-sensing data to identify crop types and estimate irrigated acreage, Uvalde and Medina Counties, 1989: U.S. Geological Survey Water-Resources Investigations Report 92–4117, 21 p.

Seaber, P.R., Kapinos, F.P., and Knapp, G.L., 1987, Hydrologic unit maps: U.S. Geological Survey Water-Supply Paper 2294, 63 p.

Solley, W.B., Chase, E.B., and Mann, W.B., IV, 1983, Estimated use of water in the United States in 1980: U.S. Geological Survey Circular 1001, 56 p.

Solley, W.B., Merk, C.F., and Pierce, R.R., 1988, Estimated use of water in the United States in 1985: U.S. Geological Survey Circular 1004, 82 p.

Solley, W.B., Pierce, R.R., and Perlman, H.A., 1993, Estimated use of water in the United States in 1990: U.S. Geological Survey Circular 1081, 76 p.

Solley, W.B., Pierce, R.R., and Perlman, H.A., 1998, Estimated use of water in the United States in 1995: U.S. Geological Survey Circular 1200, 71 p.

Stickney, R.R., 1994, Principles of aquaculture: New York, Wiley, 502 p.

Sun, R.J., Weeks, J.B., and Grubb, H.F., 1997, Bibliography of Regional Aquifer-Systems Analysis Program of the U.S. Geological Survey, 1978–96: U.S. Geological Survey Water-Resources Investigations Report 97–4074, report available on World Wide Web, accessed June 17, 2004, at *http://water.usgs.gov/ogw/rasa/html/introduction.html*

Templin, W.E., and Cherry, D.E., 1997, Drainage-return, surface water withdrawal, and land-use data for the Sacramento-San Joaquin Delta, with emphasis on Twitchell Island, California: U.S. Geological Survey Open-File Report 97–350, 31 p.

U.S. Census Bureau, 1999, 1997 Economic census—industry series (mining, construction, and manufacturing): Information available on World Wide Web, accessed April 19, 2004, at *http://www.census.gov/epcd/www/econ97.html*

U.S. Census Bureau, 2000, Bridge between NAICS and SIC—1997 Economic census, core business statistics series: Washington, D.C., 331 p.

U.S. Census Bureau, 2002, North American Industry Classification System—United States, 2002: Springfield, Virginia, 1419 p.

U.S. Congress, 1995, Paper Reduction Act of 1995: Amendments to chapter 35 of title 44, United States Code.

U.S. Department of Agriculture, 1970, Irrigation water requirements: Washington, D.C., Technical Release 21 (revision 2 of 1967 edition, available from National Technical Information Service, Springfield, Virginia, as NTIS Report PB 85–178390/XAB.), 88 p.

U.S. Department of Agriculture, 1976, Crop consumptive irrigation requirements and irrigation efficiency coefficients for the United States: Washington, D.C., SCS Special Projects Division, 118 p.

U.S. Department of Agriculture, 1997, National engineering handbook, part 652, irrigation guide: Natural Resources Conservation Service, information available on World Wide Web, accessed April 14, 2004, at *http://www.wcc.nrcs.usda.gov/nrcsirrig/irrig-handbooks-part652.html*

U.S. Department of Agriculture, 1999a, 1997 Census of agriculture: National Agricultural Statistics Service, information available on World Wide Web, accessed April 14, 2004, at *http://www.nass.usda.gov/census/census97/volume1/vol1pubs.htm*

U.S. Department of Agriculture, 1999b, 1997 Natural resources inventory: Natural Resources Conservation Service, information available on World Wide Web, accessed April 14, 2004, at *http://www.nrcs.usda.gov/technical/NRI/1997/national_results.html*

U.S. Department of Agriculture, 2000a, 1998 Census of aquaculture: National Agricultural Statistics Service, information available on World Wide Web, accessed April 16, 2004, at *http://www.nass.usda.gov/census/census97/aquaculture/aquaculture.htm*

U.S. Department of Agriculture, 2000b, 1998 Census of horticultural specialties: National Agricultural Statistics Service, information available on World Wide Web, accessed April 14, 2004, at *http://www.nass.usda.gov/census/census97/horticulture/horticulture.htm*

U.S. Department of Agriculture, 2000c, 1998 Farm and ranch irrigation survey: National Agricultural Statistics Service, information available on World Wide Web, accessed April 14, 2004, at *http://www.nass.usda.gov/census/census97/fris/fris.htm*

U.S. Department of Agriculture, 2002, 2000 Published estimates database: National Agricultural Statistics Service, information available on World Wide Web, accessed April 14, 2004, at *http://www.nass.usda.gov:81/ipedb/*

U.S. Environmental Protection Agency, 2003, Water discharge permits, permit compliance system: Information available on World Wide Web, accessed April 13, 2004, at *http://www.epa.gov/enviro/html/pcs/pcs_query_java.html*

U.S. Geological Survey, 2003, Principal aquifers, *in* National Atlas of the United States of America: Washington, D.C., 1 sheet, information available on World Wide Web, accessed June 17, 2004, at *http://nationalatlas.gov/aquifermapwhole.html*

U.S. Geological Survey, 2004, State minerals statistics and information: Information available on World Wide Web, accessed April 14, 2004, at *http://minerals.usgs.gov/minerals/pubs/state/*

van der Leeden, Frits, Troise, F.L., and Todd, D.K., 1990, The water encyclopedia (2d ed.): Chelsea, Michigan, Lewis Publishers, 808 p.

Vickers, A.L., 2001, Handbook of water use and conservation: Amherst, Massachusetts, WaterPlow Press, 446 p.

Supplemental Information

Table 1. Mandatory and optional data elements for the 2000 national water-use compilation.

[x, applies; --, does not apply]

Category and data elements	Mandatory at county level	Mandatory at State level only	Mandatory at aquifer level	Mandatory for selected States	Optional at county level[1]
Public supply					
Ground-water withdrawals - freshwater	x	--	x	--	--
Surface-water withdrawals - freshwater	x	--	--	--	--
Ground-water withdrawals - saline water	--	--	--	--	x
Surface-water withdrawals - saline water	--	--	--	--	x
Population served - total	--	x	--	--	--
Population served - ground water	--	--	--	--	x
Population served - surface water	--	--	--	--	x
Number of facilities	--	--	--	--	x
Reclaimed wastewater	--	--	--	--	x
Domestic					
Ground-water withdrawals - freshwater	--	x	--	--	x
Surface-water withdrawals - freshwater	--	x	--	--	x
Deliveries from public supply	--	--	--	--	x
Consumptive use - freshwater	--	--	--	--	x
Industrial					
Ground-water withdrawals - freshwater	x	--	x	--	--
Surface-water withdrawals - freshwater	x	--	--	--	--
Ground-water withdrawals - saline water	x	--	--	--	--
Surface-water withdrawals - saline water	x	--	--	--	--
Deliveries from public supply	--	--	--	--	x
Consumptive use - freshwater	--	--	--	--	x
Consumptive use - saline water	--	--	--	--	x
Number of facilities	--	--	--	--	x
Reclaimed wastewater	--	--	--	--	x
Irrigation					
Ground-water withdrawals - freshwater	x	--	x	--	--
Surface-water withdrawals - freshwater	x	--	--	--	--
Ground-water withdrawals - freshwater, crop only	x^2	--	x^2	--	x^2
Surface-water withdrawals - freshwater, crop only	x^2	--		--	x^2
Ground-water withdrawals - freshwater, golf course	x^2	--	x^2	--	x^2
Surface-water withdrawals - freshwater, golf course	x^2	--	--	--	x^2
Acres irrigated - sprinkler	x	--	--	--	--
Acres irrigated - surface systems	x	--	--	--	--
Acres irrigated - microirrigation	x	--	--	--	--
Consumptive use - freshwater	--	--	--	--	--
Conveyance loss	--	--	--	--	x
Reclaimed wastewater	--	--	--	--	x
Thermoelectric power generation					
Ground-water withdrawals, once through - freshwater	x	--	--	--	--
Surface-water withdrawals, once through, freshwater	x	--	--	--	--
Surface-water withdrawals, once through - saline water	x	--	--	--	--
Ground-water withdrawals, closed loop - freshwater	x	--	--	--	--
Surface-water withdrawals, closed loop - freshwater	x	--	--	--	--
Surface-water withdrawals, closed loop - saline water	x	--	--	--	--
Deliveries from public supply - once through	--	--	--	--	x

Table 1. Mandatory and optional data elements for the 2000 national water-use compilation.—Continued

[x, applies; --, does not apply]

Category and data elements	Mandatory at county level	Mandatory at State level only	Mandatory at aquifer level	Mandatory for selected States	Optional at county level[1]
Thermoelectric power generation—Continued					
Deliveries from public supply - closed loop	--	--	--	--	x
Consumptive use, once though - freshwater	--	--	--	--	x
Consumptive use, once through - saline water	--	--	--	--	x
Consumptive use, closed loop - freshwater	--	--	--	--	x
Consumptive use, closed loop - saline water	--	--	--	--	x
Power generated - once-though	--	--	--	--	x
Power generated - closed loop	--	--	--	--	x
Number of facilities - once through	--	--	--	--	x
Number of facilities - closed loop	--	--	--	--	x
Reclaimed wastewater - once through	--	--	--	--	x
Reclaimed wastewater - closed loop	--	--	--	--	x
Mining	--	--	--		
Ground-water withdrawals - freshwater	--	--	--	x^3	x
Surface-water withdrawals - freshwater	--	--	--	x^3	x
Ground-water withdrawals - saline water	--	--	--	x^3	x
Surface-water withdrawals - saline water	--	--	--	x^3	x
Consumptive use - freshwater	--	--	--	--	x
Consumptive use - saline water	--	--	--	--	x
Reclaimed wastewater	--	--	--	--	x
Livestock					
Ground-water withdrawals - freshwater	--	--	--	x^4	x
Surface-water withdrawals - freshwater	--	--	--	x^4	x
Consumptive use - freshwater	--	--	--	--	x
Aquaculture					
Ground-water withdrawals - freshwater	--	--	--	x^5	x
Surface-water withdrawals - freshwater	--	--	--	x^5	x
Ground-water withdrawals - saline water	--	--	--	--	x
Surface-water withdrawals - saline water	--	--	--	--	x
Consumptive use - freshwater	--	--	--	--	x
Consumptive use - saline water	--	--	--	--	x
Commercial					
Ground-water withdrawals - freshwater	--	--	--	--	x
Surface-water withdrawals - freshwater	--	--	--	--	x
Deliveries from public supply	--	--	--	--	x
Consumptive use - freshwater	--	--	--	--	x
Hydroelectric power generation					
Instream water use	--	--	--	--	x
Power generation	--	--	--	--	x
Number of facilities	--	--	--	--	x

Table 1. Mandatory and optional data elements for the 2000 national water-use compilation.—Continued

[x, applies; --, does not apply]

Category and data elements	Mandatory at county level	Mandatory at State level only	Mandatory at aquifer level	Mandatory for selected States	Optional at county level[1]
Wastewater treatment					
Number of facilities	--	--	--	--	x
Wastewater returns - public facilities	--	--	--	--	x
Reclaimed wastewater	--	--	--	--	x

[1]Any data aggregated by hydrologic cataloging unit are optional for all States.

[2]Subdivision of irrigation withdrawals into crop and golf-course categories is optional, but both categories are mandatory if withdrawals are subdivided.

[3]Mining withdrawals are mandatory for Alaska, Arizona, California, Florida, Indiana, Minnesota, Nebraska, Oklahoma, Pennsylvania, Texas, Utah, and Wyoming, and are optional for all other States.

[4]Livestock withdrawals are mandatory for California, Iowa, Kansas, Minnesota, Missouri, Nebraska, North Carolina, Oklahoma, Texas, and Wisconsin, and are optional for all other States.

[5]Aquaculture withdrawals are mandatory for Alabama, Arkansas, California, Idaho, Louisiana, Mississippi, North Carolina, and Utah, and are optional for all other States.

Table 2. List of principal aquifers and aquifer systems by State that are used by the U.S. Geological Survey for aggregation of 2000 water-use data.

State	Name of principal aquifer or aquifer system
Alabama	Coastal lowlands aquifer system
	Floridan aquifer system
	Mississippi embayment aquifer system
	Mississippian aquifers
	Pennsylvanian aquifers
	Piedmont and Blue Ridge crystalline-rock aquifers
	Southeastern Coastal Plain aquifer system
	Surficial aquifer system
	Valley and Ridge aquifers
	Valley and Ridge carbonate-rock aquifers
Alaska	Unconsolidated-deposit aquifers
Arizona	Basin and Range basin-fill aquifers
	Basin and Range carbonate-rock aquifers
	Colorado Plateaus aquifers
Arkansas	Alluvial aquifers
	Edwards-Trinity aquifer system
	Mississippi embayment aquifer system
	Mississippi River Valley alluvial aquifer
	Ozark Plateaus aquifer system
California	Basin and Range basin-fill aquifers
	Basin and Range carbonate-rock aquifers
	California Coastal Basin aquifers
	Central Valley aquifer system
	Pacific Northwest basin-fill aquifers
	Pacific Northwest volcanic-rock aquifers
Colorado	Alluvial aquifers
	Colorado Plateaus aquifers
	Denver Basin aquifer system
	High Plains aquifer
	Rio Grande aquifer system
Connecticut	Early Mesozoic basin aquifers
	New England crystalline-rock aquifers
	New York and New England carbonate-rock aquifers
	Sand and gravel aquifers (glaciated regions)
Delaware	Northern Atlantic Coastal Plain aquifer system
	Piedmont and Blue Ridge crystalline-rock aquifers
	Surficial aquifer system
District of Columbia	Northern Atlantic Coastal Plain aquifer system
	Piedmont and Blue Ridge crystalline-rock aquifers

Table 2. List of principal aquifers and aquifer systems by State that are used by the U.S. Geological Survey for aggregation of 2000 water-use data.—Continued

State	Name of principal aquifer or aquifer system
Florida	Biscayne aquifer
	Coastal lowlands aquifer system
	Floridan aquifer system
	Intermediate aquifer system
	Surficial aquifer system
Georgia	Floridan aquifer system
	Pennsylvanian aquifers
	Piedmont and Blue Ridge crystalline-rock aquifers
	Southeastern Coastal Plain aquifer system
	Surficial aquifer system
	Valley and Ridge aquifers
	Valley and Ridge carbonate-rock aquifers
Hawaii	Volcanic-rock aquifers (Hawaii)
Idaho	Basin and Range basin-fill aquifers
	Basin and Range carbonate-rock aquifers
	Columbia Plateau basaltic-rock aquifers
	Columbia Plateau basin-fill aquifers
	Northern Rocky Mountains Intermontane Basins aquifer systems
	Pacific Northwest basin-fill aquifers
	Pacific Northwest volcanic-rock aquifers
	Snake River Plain basaltic-rock aquifers
	Snake River Plain basin-fill aquifers
Illinois	Cambrian-Ordovician aquifer system
	Mississippi embayment aquifer system
	Mississippi River Valley alluvial aquifer
	Mississippian aquifers
	Ozark Plateaus aquifer system
	Pennsylvanian aquifers
	Sand and gravel aquifers (glaciated regions)
	Silurian-Devonian aquifers
	Southeastern Coastal Plain aquifer system
Indiana	Mississippian aquifers
	Pennsylvanian aquifers
	Sand and gravel aquifers (glaciated regions)
	Silurian-Devonian aquifers
Iowa	Cambrian-Ordovician aquifer system
	Lower Cretaceous aquifers
	Mississippian aquifers
	Sand and gravel aquifers (glaciated regions)
	Silurian-Devonian aquifers
	Upper Carbonate aquifer

Table 2. List of principal aquifers and aquifer systems by State that are used by the U.S. Geological Survey for aggregation of 2000 water-use data.—Continued

State	Name of principal aquifer or aquifer system
Kansas	Alluvial aquifers
	High Plains aquifer
	Lower Cretaceous aquifers
	Ozark Plateaus aquifer system
	Sand and gravel aquifers (glaciated regions)
Kentucky	Alluvial aquifers
	Mississippi embayment aquifer system
	Mississippi River Valley alluvial aquifer
	Mississippian aquifers
	Ordovician aquifers
	Pennsylvanian aquifers
	Sand and gravel aquifers (glaciated regions)
	Silurian-Devonian aquifers
Louisiana	Alluvial aquifers
	Coastal lowlands aquifer system
	Mississippi embayment aquifer system
	Mississippi River Valley alluvial aquifer
	Texas coastal uplands aquifer system
Maine	New England crystalline-rock aquifers
	New York and New England carbonate-rock aquifers
	Sand and gravel aquifers (glaciated regions)
Maryland	Early Mesozoic basin aquifers
	Mississippian aquifers
	Northern Atlantic Coastal Plain aquifer system
	Pennsylvanian aquifers
	Piedmont and Blue Ridge carbonate-rock aquifers
	Piedmont and Blue Ridge crystalline-rock aquifers
	Surficial aquifer system
	Valley and Ridge aquifers
	Valley and Ridge carbonate-rock aquifers
Massachusetts	Early Mesozoic basin aquifers
	New England crystalline-rock aquifers
	New York and New England carbonate-rock aquifers
	Sand and gravel aquifers (glaciated regions)
Michigan	Cambrian-Ordovician aquifer system
	Jacobsville aquifer
	Marshall aquifer
	Pennsylvanian aquifers
	Sand and gravel aquifers (glaciated regions)
	Silurian-Devonian aquifers

Table 2. List of principal aquifers and aquifer systems by State that are used by the U.S. Geological Survey for aggregation of 2000 water-use data.—Continued

State	Name of principal aquifer or aquifer system
Minnesota	Cambrian-Ordovician aquifer system
	Lower Cretaceous aquifers
	Paleozoic aquifers
	Sand and gravel aquifers (glaciated regions)
	Upper Carbonate aquifer
Mississippi	Coastal lowlands aquifer system
	Mississippi embayment aquifer system
	Mississippi River Valley alluvial aquifer
	Southeastern Coastal Plain aquifer system
Missouri	Alluvial aquifers
	Cambrian-Ordovician aquifer system
	Mississippi embayment aquifer system
	Mississippi River Valley alluvial aquifer
	Mississippian aquifers
	Ozark Plateaus aquifer system
	Sand and gravel aquifers (glaciated regions)
	Silurian-Devonian aquifers
Montana	Alluvial aquifers
	Lower Cretaceous aquifers
	Lower Tertiary aquifers
	Northern Great Plains aquifer system
	Northern Rocky Mountains Intermontane Basins aquifer systems
	Pacific Northwest volcanic-rock aquifers
	Paleozoic aquifers
	Sand and gravel aquifers (glaciated regions)
	Upper Cretaceous aquifers
Nebraska	Alluvial aquifers
	High Plains aquifer
	Lower Cretaceous aquifers
	Sand and gravel aquifers (glaciated regions)
Nevada	Basin and Range basin-fill aquifers
	Basin and Range carbonate-rock aquifers
	Pacific Northwest basin-fill aquifers
	Pacific Northwest volcanic-rock aquifers
	Southern Nevada volcanic-rock aquifers
New Hampshire	New England crystalline-rock aquifers
	Sand and gravel aquifers (glaciated regions)
New Jersey	Early Mesozoic basin aquifers
	New York and New England carbonate-rock aquifers
	Northern Atlantic Coastal Plain aquifer system
	Piedmont and Blue Ridge carbonate-rock aquifers

Table 2. List of principal aquifers and aquifer systems by State that are used by the U.S. Geological Survey for aggregation of 2000 water-use data.—Continued

State	Name of principal aquifer or aquifer system
New Jersey—Continued	Piedmont and Blue Ridge crystalline-rock aquifers
	Sand and gravel aquifers (glaciated regions)
	Surficial aquifer system
	Valley and Ridge aquifers
	Valley and Ridge carbonate-rock aquifers
New Mexico	Basin and Range basin-fill aquifers
	Colorado Plateaus aquifers
	High Plains aquifer
	Pecos River Basin alluvial aquifer
	Rio Grande aquifer system
	Roswell Basin aquifer system
New York	Early Mesozoic basin aquifers
	New England crystalline-rock aquifers
	New York and New England carbonate-rock aquifers
	New York Sandstone aquifers
	Northern Atlantic Coastal Plain aquifer system
	Piedmont and Blue Ridge crystalline-rock aquifers
	Sand and gravel aquifers (glaciated regions)
	Valley and Ridge aquifers
North Carolina	Castle Hayne aquifer
	Early Mesozoic basin aquifers
	Northern Atlantic Coastal Plain aquifer system
	Piedmont and Blue Ridge carbonate-rock aquifers
	Piedmont and Blue Ridge crystalline-rock aquifers
	Southeastern Coastal Plain aquifer system
	Surficial aquifer system
	Valley and Ridge aquifers
North Dakota	Alluvial aquifers
	Lower Cretaceous aquifers
	Lower Tertiary aquifers
	Northern Great Plains aquifer system
	Paleozoic aquifers
	Sand and gravel aquifers (glaciated regions)
	Upper Cretaceous aquifers
Ohio	Alluvial aquifers
	Mississippian aquifers
	Pennsylvanian aquifers
	Sand and gravel aquifers (glaciated regions)
	Silurian-Devonian aquifers
Oklahoma	Ada-Vamoosa aquifer
	Alluvial aquifers
	Arbuckle-Simpson aquifer

Table 2. List of principal aquifers and aquifer systems by State that are used by the U.S. Geological Survey for aggregation of 2000 water-use data.—Continued

State	Name of principal aquifer or aquifer system
Oklahoma—Continued	Blaine aquifer
	Central Oklahoma aquifer
	Edwards-Trinity aquifer system
	High Plains aquifer
	Ozark Plateaus aquifer system
	Rush Springs aquifer
Oregon	Basin and Range basin-fill aquifers
	Columbia Plateau basaltic-rock aquifers
	Columbia Plateau basin-fill aquifers
	Pacific Northwest basin-fill aquifers
	Pacific Northwest volcanic-rock aquifers
	Puget Sound aquifer system
	Snake River Plain basaltic-rock aquifers
	Snake River Plain basin-fill aquifers
Pennsylvania	Early Mesozoic basin aquifers
	Mississippian aquifers
	New York and New England carbonate-rock aquifers
	Northern Atlantic Coastal Plain aquifer system
	Pennsylvanian aquifers
	Piedmont and Blue Ridge carbonate-rock aquifers
	Piedmont and Blue Ridge crystalline-rock aquifers
	Sand and gravel aquifers (glaciated regions)
	Valley and Ridge aquifers
	Valley and Ridge carbonate-rock aquifers
Puerto Rico	North Coast Limestone aquifer system (Puerto Rico)
	South Coast aquifer (Puerto Rico)
Rhode Island	New England crystalline-rock aquifers
	Sand and gravel aquifers (glaciated regions)
South Carolina	Floridan aquifer system
	Piedmont and Blue Ridge crystalline-rock aquifers
	Southeastern Coastal Plain aquifer system
	Surficial aquifer system
South Dakota	Alluvial aquifers
	High Plains aquifer
	Lower Cretaceous aquifers
	Lower Tertiary aquifers
	Northern Great Plains aquifer system
	Paleozoic aquifers
	Sand and gravel aquifers (glaciated regions)
	Upper Cretaceous aquifers

Table 2. List of principal aquifers and aquifer systems by State that are used by the U.S. Geological Survey for aggregation of 2000 water-use data.—Continued

State	Name of principal aquifer or aquifer system
Tennessee	Alluvial aquifers
	Mississippi embayment aquifer system
	Mississippi River Valley alluvial aquifer
	Mississippian aquifers
	Ordovician aquifers
	Pennsylvanian aquifers
	Piedmont and Blue Ridge crystalline-rock aquifers
	Silurian-Devonian aquifers
	Southeastern Coastal Plain aquifer system
	Valley and Ridge aquifers
	Valley and Ridge carbonate-rock aquifers
Texas	Alluvial aquifers
	Blaine aquifer
	Coastal lowlands aquifer system
	Edwards-Trinity aquifer system
	High Plains aquifer
	Mississippi embayment aquifer system
	Pecos River Basin alluvial aquifer
	Rio Grande aquifer system
	Seymour aquifer
	Texas coastal uplands aquifer system
Utah	Alluvial aquifers
	Basin and Range basin-fill aquifers
	Basin and Range carbonate-rock aquifers
	Colorado Plateaus aquifers
	Pacific Northwest basin-fill aquifers
	Pacific Northwest volcanic-rock aquifers
Vermont	New England crystalline-rock aquifers
	New York and New England carbonate-rock aquifers
	Sand and gravel aquifers (glaciated regions)
Virgin Islands	Kingshill aquifer (Virgin Islands)
Virginia	Early Mesozoic basin aquifers
	Mississippian aquifers
	Northern Atlantic Coastal Plain aquifer system
	Pennsylvanian aquifers
	Piedmont and Blue Ridge crystalline-rock aquifers
	Surficial aquifer system
	Valley and Ridge aquifers
	Valley and Ridge carbonate-rock aquifers
Washington	Columbia Plateau basaltic-rock aquifers
	Columbia Plateau basin-fill aquifers
	Northern Rocky Mountains Intermontane Basins aquifer systems

Table 2. List of principal aquifers and aquifer systems by State that are used by the U.S. Geological Survey for aggregation of 2000 water-use data.—Continued

State	Name of principal aquifer or aquifer system
Washington—Continued	Pacific Northwest basin-fill aquifers
	Pacific Northwest volcanic-rock aquifers
	Puget Sound aquifer system
	Willamette Lowland basin-fill aquifers
West Virginia	Alluvial aquifers
	Mississippian aquifers
	Pennsylvanian aquifers
	Valley and Ridge aquifers
	Valley and Ridge carbonate-rock aquifers
Wisconsin	Cambrian-Ordovician aquifer system
	Jacobsville aquifer
	Sand and gravel aquifers (glaciated regions)
	Silurian-Devonian aquifers
Wyoming	Alluvial aquifers
	Colorado Plateaus aquifers
	High Plains aquifer
	Lower Cretaceous aquifers
	Lower Tertiary aquifers
	Northern Great Plains aquifer system
	Northern Rocky Mountains Intermontane Basins aquifer systems
	Pacific Northwest basin-fill aquifers
	Pacific Northwest volcanic-rock aquifers
	Paleozoic aquifers
	Upper Cretaceous aquifers
	Wyoming Tertiary aquifers

Table 3. List of Standard Industrial Classification (SIC) codes by water-use category.

Water-use category	SIC code or range	Description
Public supply	4941	Water supply
Domestic	8811	Private households
Industrial	1521–1522	General building contractors—residential buildings
	1531	Operative builders
	1541–1542	General building contractors—nonresidential buildings
	1611	Highway and street construction, except elevated highways
	1622–1629	Heavy construction, except highway and street construction
	1711	Plumbing, heating, air-conditioning
	1721	Painting and paper hanging
	1731	Electrical work
	1741–1743	Masonry, stonework, tile setting, and plastering
	1751–1752	Carpentry and floor work
	1761	Roofing, siding, and sheet metal work
	1771	Concrete work
	1781	Water well drilling
	1791–1799	Miscellaneous special trade contractors
	2011–2015	Meat products
	2021–2026	Dairy products
	2032–2038	Canned, frozen, and preserved fruits, and vegetables, and food specialties
	2041–2048	Grain mill products
	2051–2053	Bakery products
	2061–2068	Sugar and confectionery products
	2074–2079	Fats and oils
	2082–2087	Beverages
	2091–2099	Miscellaneous food preparations and kindred products
	2111	Cigarettes
	2121	Cigars
	2131	Chewing and smoking tobacco and snuff
	2141	Tobacco stemming and redrying
	2211	Broadwoven fabric mills, cotton
	2221	Broadwoven fabric mills, manmade fiber and silk
	2231	Broadwoven fabric mills, wool (including dyeing and finishing)
	2241	Narrow fabric and other smallwares mills: cotton, wool, silk, and manmade fiber
	2251–2259	Knitting mills
	2261–2269	Dyeing and finishing textiles, except wool fabrics and knit goods
	2273	Carpets and rugs
	2281–2284	Yarn and thread mills
	2295–2299	Miscellaneous textile goods

Table 3. List of Standard Industrial Classification (SIC) codes by water-use category.—Continued

Water-use category	SIC code or range	Description
Industrial—Continued	2311	Men's and boys' suits, coats and overcoats
	2321–2329	Men's and boys' furnishings, work clothing, and allied garments
	2331–2339	Women's, misses', and juniors' outerwear
	2341–2342	Women's, misses', children's, and infants' undergarments
	2353	Hats, caps, and millinery
	2361–2369	Girls', children's, and infants' outerwear
	2371	Fur goods
	2381–2389	Miscellaneous apparel and accessories
	2391–2399	Miscellaneous fabricated textile products
	2411	Logging
	2421–2429	Sawmills and planing mills
	2431–2439	Millwork, veneer, plywood, and structural wood members
	2441–2449	Wood containers
	2451–2452	Wood buildings and mobile homes
	2491–2499	Miscellaneous wood products
	2511–2519	Household furniture
	2521–2522	Office furniture
	2531	Public building and related furniture
	2541–2542	Partitions, shelving, lockers, and office and store fixtures
	2591–2599	Miscellaneous furniture and fixtures
	2611	Pulp mills
	2621	Paper mills
	2631	Paperboard mills
	2652–2657	Paperboard containers and boxes
	2671–2679	Converted paper and paperboard products, except containers and boxes
	2711	Newspapers: publishing, or publishing and printing
	2721	Periodicals: publishing, or publishing and printing
	2731–2732	Books
	2741	Miscellaneous publishing
	2752–2759	Commercial printing
	2761	Manifold business forms
	2771	Greeting cards
	2782–2789	Blankbooks, looseleaf binders, and bookbinding and related work
	2791–2796	Service industries for the printing trade
	2812–2819	Industrial inorganic chemicals
	2821–2824	Plastics materials and synthetic resins, synthetic rubber, cellulosic and other manmade fibers, except glass
	2833–2836	Drugs
	2841–2844	Soap, detergents, and cleaning preparations; perfumes, cosmetics, and other toilet preparations
	2851	Paints, varnishes, lacquers, enamels, and allied products
	2861–2869	Industrial organic chemicals

Table 3. List of Standard Industrial Classification (SIC) codes by water-use category.—Continued

Water-use category	SIC code or range	Description
Industrial—Continued	2873–2879	Agricultural chemicals
	2891–2899	Miscellaneous chemical products
	2911	Petroleum refining
	2951–2952	Asphalt paving and roofing materials
	2992–2999	Miscellaneous products of petroleum and coal
	3011	Tires and inner tubes
	3021	Rubber and plastics footwear
	3052–3053	Gaskets, packing, and sealing devices and rubber and plastic hose and belting
	3061–3069	Fabricated rubber products, not elsewhere classified
	3081–3089	Miscellaneous plastics products
	3111	Leather tanning and finishing
	3131	Boot and shoe cut stock and findings
	3142–3149	Footwear, except rubber
	3151	Leather gloves and mittens
	3161	Luggage
	3171–3172	Handbags and other personal leather goods
	3199	Leather goods, not elsewhere classified
	3211	Flat glass
	3221–3229	Glass and glassware, pressed or blown
	3231	Glass products, made of purchased glass
	3241	Cement, hydraulic
	3251–3259	Structural clay products
	3261–3269	Pottery and related products
	3271–3275	Concrete, gypsum, and plaster products
	3281	Cut stone and stone products
	3291–3299	Abrasive, asbestos, and miscellaneous nonmetallic mineral products
	3312–3317	Steel works, blast furnaces, and rolling and finishing mills
	3321–3325	Iron and steel foundries
	3331–3339	Primary smelting and refining of nonferrous metals
	3341	Secondary smelting and refining of nonferrous metals
	3351–3357	Rolling, drawing, and extruding of nonferrous metals
	3363–3369	Nonferrous foundries (castings)
	3398–3399	Miscellaneous primary metal products
	3411–3412	Metal cans and shipping containers
	3421–3429	Cutlery, handtools, and general hardware
	3431–3433	Heating equipment, except electric and warm air; and plumbing fixtures
	3441–3449	Fabricated structural metal products
	3451–3452	Screw machine products, and bolts, nuts, screws, rivets, and washers
	3462–3469	Metal forgings and stampings
	3471–3479	Coating, engraving, and allied services

Table 3. List of Standard Industrial Classification (SIC) codes by water-use category.—Continued

Water-use category	SIC code or range	Description
Industrial—Continued	3482–3489	Ordnance and accessories, except vehicles and guided missiles
	3491–3499	Miscellaneous fabricated metal products
	3511–3519	Engines and turbines
	3523–3524	Farm and garden machinery and equipment
	3531–3537	Construction, mining, and materials handling machinery and equipment
	3541–3549	Metalworking machinery and equipment
	3552–3559	Special industry machinery, except metalworking machinery
	3561–3569	General industrial machinery and equipment
	3571–3579	Computer and office equipment
	3581–3589	Refrigeration and service industry machinery
	3592–3599	Miscellaneous industrial and commercial machinery and equipment
	3612–3613	Electric transmission and distribution equipment
	3621–3629	Electrical industrial apparatus
	3631–3639	Household appliances
	3641–3648	Electric lighting and wiring equipment
	3651–3652	Household audio and video equipment, and audio recordings
	3661–3669	Communications equipment
	3671–3679	Electronic components and accessories
	3691–3699	Miscellaneous electrical machinery, equipment, and supplies
	3711–3716	Motor vehicles and motor vehicle equipment
	3721–3728	Aircraft and parts
	3731–3732	Ship and boat building and repairing
	3743	Railroad equipment
	3751	Motorcycles, bicycles, and parts
	3761–3769	Guided missiles and space vehicles and parts
	3792–3799	Miscellaneous transportation equipment
	3812	Search, detection, navigation, guidance, aeronautical, and nautical systems, instruments, and equipment
	3821–3829	Laboratory apparatus and analytical, optical, measuring and controlling instruments
	3841–3845	Surgical, medical, and dental instruments and supplies
	3851	Ophthalmic goods
	3861	Photographic equipment and supplies
	3873	Watches, clocks, clockwork operated devices, and parts
	3911–3915	Jewelry, silverware, and plated ware
	3931	Musical instruments
	3942–3949	Dolls, toys, games and sporting and athletic goods
	3951–3955	Pens, pencils, and other artists' materials
	3961–3965	Costume jewelry, costume novelties, buttons, and miscellaneous notions, except precious metal
	3991–3999	Miscellaneous manufacturing industries
Irrigation	0111–0119	Cash grains (wheat, rice, corn, soybeans, other)

Table 3. List of Standard Industrial Classification (SIC) codes by water-use category.—Continued

Water-use category	SIC code or range	Description
Irrigation—Continued	0131–0139	Field crops, except cash grains (cotton, tobacco, sugarcane, sugarbeets, potatoes, alfalfa, hay, other)
	0161	Vegetables and melons
	0171–0179	Fruits and tree nuts (berry crops, grapes, tree nuts, citrus fruit, deciduous tree fruits, other)
	0181–0182	Horticultural specialties (ornamental floriculture and nursery products including sod farms, food crops grown under cover)
	0191	General farms, primarily crop
	4971	Irrigation systems
	7992	Public golf courses
	7997	Membership sports and recreation clubs
Thermoelectric power	4911	Electric services
Mining	1011	Iron ores
	1021	Copper ores
	1031	Lead and zinc ores
	1041–1044	Gold and silver ores
	1061	Ferroalloy ores, except vanadium
	1081	Metal mining services
	1094–1099	Miscellaneous metal ores
	1221–1222	Bituminous coal and lignite mining
	1231	Anthracite mining
	1241	Coal mining services
	1311	Crude petroleum and natural gas
	1321	Natural gas liquids
	1381–1389	Oil and gas field services
	1411	Dimension stone
	1422–1429	Crushed and broken stone, including riprap
	1442–1446	Sand and gravel
	1455–1459	Clay, ceramic, and refractory minerals
	1474–1479	Chemical and fertilizer mineral mining
	1481	Nonmetallic minerals services, except fuels
	1499	Miscellaneous nonmetallic minerals, except fuels
Livestock	0211–0219	Livestock, except dairy and poultry (beef cattle, hogs, sheep and goats)
	0240–0241	Dairy farms
	0251–0259	Poultry and eggs farms (chickens, turkeys, ducks, other)
	0271	Fur-bearing animals and rabbits
	0272	Horses and other equines
	0279	Animal specialties, not elsewhere classified
	0291	General farms, primarily livestock and animal specialties
Aquaculture	0273	Animal aquaculture (catfish, crustaceans, finfish, tropical fish, trout)
	0921	Fish hatcheries and preserves

Table 3. List of Standard Industrial Classification (SIC) codes by water-use category.—Continued

Water-use category	SIC code or range	Description
Commercial	0711	Soil preparation services
	0721	Crop services
	0741	Veterinary services
	0751	Animal services, except veterinary
	0761–0762	Farm labor and management services
	0781–0783	Landscape horticultural services
	0811	Timber tracts
	0831	Forest nurseries and gathering of forest products
	0851	Forestry services
	0912–0919	Commercial fishing
	0971	Hunting and trapping, and game propagation
	4011–4013	Railroads
	4111–4119	Local and suburban transportation
	4121	Taxicabs
	4131	Intercity and rural bus transportation
	4141–4142	Bus charter service
	4151	School buses
	4173	Terminal and service facilities for motor vehicle passenger transportation
	4212–4215	Trucking and courier services, except air
	4221–4226	Public warehousing and storage
	4231	Terminal and joint terminal maintenance facilities for motor freight transportation
	4311	U.S. Postal Service
	4412	Deep sea foreign transportation of freight
	4424	Deep sea domestic transportation of freight
	4432	Freight transportation on the Great Lakes—St. Lawrence Seaway
	4449	Water transportation of freight, not elsewhere classified
	4481–4489	Water transportation of passengers
	4491–4499	Services incidental to water transportation
	4512–4513	Air transportation, scheduled, and air courier services
	4522	Air transportation, nonscheduled
	4581	Airports, flying fields, and airport terminal services
	4612–4619	Pipelines, except natural gas
	4724–4729	Arrangement of passenger transportation
	4731	Arrangement of transportation of freight and cargo
	4741	Rental of railroad cars
	4783–4789	Miscellaneous services incidental to transportation
	4812–4813	Telephone communications
	4822	Telegraph and other message communications
	4832–4833	Radio and television broadcasting stations
	4841	Cable and other pay television services

Table 3. List of Standard Industrial Classification (SIC) codes by water-use category.—Continued

Water-use category	SIC code or range	Description
Commercial—Continued	4899	Communications services, not elsewhere classified
	4922–4925	Gas production and distribution
	4931–4939	Combination electric and gas, other utility services
	4953–4959	Sanitary services, other than sewage systems
	4961	Steam and air-conditioning supply
	5012–5015	Motor vehicles and motor vehicle parts and supplies
	5021–5023	Furniture and home furnishings
	5031–5039	Lumber and construction materials
	5043–5049	Professional and commercial, equipment and supplies
	5051–5052	Metals and minerals, except petroleum
	5063–5065	Electrical goods
	5072–5078	Hardware, and plumbing and heating equipment and supplies
	5082–5088	Machinery, equipment, and supplies
	5091–5099	Miscellaneous durable goods
	5111–5113	Paper and paper products
	5122	Drugs, drug proprietaries, and druggists' sundries
	5131–5139	Apparel, piece goods, and notions
	5141–5149	Groceries and related products
	5153–5159	Farm-product raw materials
	5162–5169	Chemicals and allied products
	5171–5172	Petroleum and petroleum products
	5181–5182	Beer, wine, and distilled alcoholic beverages
	5191–5199	Miscellaneous nondurable goods
	5211	Lumber and other building materials dealers
	5231	Paint, glass, and wallpaper stores
	5251	Hardware stores
	5261	Retail nurseries and garden supply stores
	5271	Mobile home dealers
	5311	Department stores
	5331	Variety stores
	5399	Miscellaneous general merchandise stores
	5411	Grocery stores
	5421	Meat and fish (seafood) markets, including freezer provisioners
	5431	Fruit and vegetable markets
	5441	Candy, nut, and confectionery stores
	5451	Dairy products stores
	5461	Retail bakeries
	5499	Miscellaneous food stores
	5511	Motor vehicle dealers (new and used)
	5521	Motor vehicle dealers (used only)

Table 3. List of Standard Industrial Classification (SIC) codes by water-use category.—Continued

Water-use category	SIC code or range	Description
Commercial—Continued	5531	Auto and home supply stores
	5541	Gasoline service stations
	5551	Boat dealers
	5561	Recreational vehicle dealers
	5571	Motorcycle dealers
	5599	Automotive dealers, not elsewhere classified
	5611	Men's and boys' clothing and accessory stores
	5621	Women's clothing stores
	5632	Women's accessory and specialty stores
	5641	Children's and infants' wear stores
	5651	Family clothing stores
	5661	Shoe stores
	5699	Miscellaneous apparel and accessory stores
	5712–5719	Home furniture and furnishing stores
	5722	Household appliance stores
	5731–5736	Radio, television, consumer electronics, and music stores
	5812–5813	Eating and drinking places
	5912	Drug stores and proprietary stores
	5921	Liquor stores
	5932	Used merchandise stores
	5941–5949	Miscellaneous shopping goods stores
	5961–5963	Nonstore retailers
	5983–5989	Fuel dealers
	5992–5999	Retail stores, not elsewhere classified
	6011–6019	Central Reserve depository institutions
	6021–6029	Commercial banks
	6035–6036	Savings institutions
	6061–6062	Credit unions
	6081–6082	Foreign banking and branches and agencies of foreign banks
	6091–6099	Functions related to depository banking
	6111	Federal and Federally sponsored credit agencies
	6141	Personal credit institutions
	6153–6159	Business credit institutions
	6162–6163	Mortgage bankers and brokers
	6211	Security brokers, dealers, and flotation companies
	6221	Commodity contracts brokers and dealers
	6231	Security and commodity exchanges
	6182–6189	Services allied with the exchange of securities or commodities
	6311	Life insurance
	6321–6324	Accident and health insurance and medical service plans

Table 3. List of Standard Industrial Classification (SIC) codes by water-use category.—Continued

Water-use category	SIC code or range	Description
Commercial—Continued	6331	Fire, marine, and casualty insurance
	6351	Surety insurance
	6361	Title insurance
	6371	Pension, health, and welfare funds
	6399	Insurance carriers, not elsewhere classified
	6411	Insurance agents, brokers, and service
	6512–6519	Real estate operators (except developers) and lessors
	6531	Real estate agents and managers
	6541	Title abstract offices
	6552–6553	Land subdividers and developers
	6712–6719	Holding offices
	6722–6726	Investment offices
	6732–6733	Trusts
	6792–6799	Miscellaneous investing
	7011	Hotels and motels
	7021	Rooming and boarding houses
	7032–7033	Camps and recreational vehicle parks
	7041	Organization hotels and lodging houses, on membership basis
	7211–7219	Laundry, cleaning, and garment services
	7221	Photographic studios, portrait
	7231	Beauty shops
	7241	Barber shops
	7251	Shoe repair shops and shoeshine parlors
	7261	Funeral service and crematories
	7291–7299	Miscellaneous personal services
	7311–7319	Advertising
	7322–7323	Consumer credit reporting agencies, mercantile reporting agencies, and adjustment and collection agencies
	7331–7338	Mailing, reproduction, commercial art and photography, and stenographic services
	7342–7349	Services to dwellings and other buildings
	7352–7359	Miscellaneous equipment rental and leasing
	7361–7363	Personnel supply services
	7371–7379	Computer programming, data processing, and other computer related services
	7381–7389	Miscellaneous business services
	7513–7519	Automotive rental and leasing, without drivers
	7521	Automobile parking
	7532–7539	Automotive repair shops
	7542–7549	Automotive services, except repair
	7622–7629	Electrical repair shops
	7631	Watch, clock, and jewelry repair
	7641	Re-upholstery and furniture repair

Table 3. List of Standard Industrial Classification (SIC) codes by water-use category.—Continued

Water-use category	SIC code or range	Description
Commercial—Continued	7692–7699	Miscellaneous repair shops and related services
	7812–7819	Motion picture production and allied services
	7822–7829	Motion picture distribution and allied services
	7832–7833	Motion picture theaters
	7841	Video tape rental
	7911	Dance studios, schools, and halls
	7922–7929	Theatrical producers (except motion picture), bands, orchestras, and entertainers
	7933	Bowling centers
	7941–7948	Commercial sports
	7991–7999	Miscellaneous amusement and recreation services
	8011	Offices and clinics of doctors of medicine
	8021	Offices and clinics of dentists
	8031	Offices and clinics of doctors of osteopathy
	8041–8049	Offices and clinics of other health practitioners
	8051–8059	Nursing and personal care facilities
	8062–8069	Hospitals
	8071–8072	Medical and dental laboratories
	8082	Home health care services
	8292–8299	Miscellaneous health and allied services, not elsewhere classified
	8111	Legal services
	8211	Elementary and secondary schools
	8221–8222	Colleges and universities
	8231	Libraries
	8243–8249	Vocational schools
	8299	Schools and educational services, not elsewhere classified
	8322	Individual and family social services
	8331	Job training and related services
	8351	Child day care services
	8361	Residential care
	8399	Social services, not elsewhere classified
	8412	Museums and art galleries
	8422	Arboreta and botanical or zoological gardens
	8611	Business associations
	8621	Professional membership organizations
	8631	Labor unions and similar labor organizations
	8641	Civic, social, and fraternal associations
	8651	Political organizations
	8661	Religious organizations
	8699	Membership organizations, not elsewhere classified
	8711–8713	Engineering, architectural, and surveying services

Table 3. List of Standard Industrial Classification (SIC) codes by water-use category.—Continued

Water-use category	SIC code or range	Description
Commercial—Continued	8721	Accounting, auditing, and bookkeeping services
	8731–8734	Research and testing services
	8741–8748	Management and public relations
	8999	Services, not elsewhere classified
	9111	Executive offices
	9121	Legislative bodies
	9131	Executive and legislative offices combined
	9199	General government, not elsewhere classified
	9211	Courts
	9221–9229	Public order and safety
	9311	Public finance, taxation, and monetary policy
	9411	Administration of educational programs
	9431	Administration of public health programs
	9441	Administration of social, human resource and income maintenance programs
	9451	Administration of veterans' affairs, except health and insurance
	9511–9512	Administration of environmental quality programs
	9531–9532	Housing and urban development
	9611	Administration of general economic programs
	9621	Regulation and administration of transportation programs
	9631	Regulation and administration of communications, electric, gas and other utilities
	9641	Regulation of agricultural marketing and commodities
	9651	Regulation, licensing, and inspection of miscellaneous commercial sectors
	9661	Space research and technology
	9711	National security
	9721	International affairs
Hydroelectric power	4911	Electric services